OFFICIAL SQA PAST PAPERS WITH ANSWERS

STANDARD GRADE | CREDIT

BIOLOGY
2006-2010

First exam published in 2006.
Published by Bright Red Publishing Ltd, 6 Stafford Street, Edinburgh EH3 7AU
tel: 0131 220 5804 fax: 0131 220 6710 info@brightredpublishing.co.uk www.brightredpublishing.co.uk

ISBN 978-1-84948-080-2

A CIP Catalogue record for this book is available from the British Library.

Bright Red Publishing is grateful to the copyright holders, as credited on the final page of the book, for permission to use their material. Every effort has been made to trace the copyright holders and to obtain their permission for the use of copyright material. Bright Red Publishing will be happy to receive information allowing us to rectify any error or omission in future editions.

STANDARD GRADE | CREDIT

2006

[BLANK PAGE]

FOR OFFICIAL USE

C

KU PS

Total Marks

0300/402

NATIONAL QUALIFICATIONS 2006

TUESDAY, 23 MAY
10.50 AM – 12.20 PM

BIOLOGY
STANDARD GRADE
Credit Level

Fill in these boxes and read what is printed below.

Full name of centre

Town

Forename(s)

Surname

Date of birth
Day Month Year

Scottish candidate number

Number of seat

1 All questions should be attempted.

2 The questions may be answered in any order but all answers are to be written in the spaces provided in this answer book, and must be written clearly and legibly in ink.

3 Rough work, if any should be necessary, as well as the fair copy, is to be written in this book. Additional spaces for answers and for rough work will be found at the end of the book. Rough work should be scored through when the fair copy has been written.

4 Before leaving the examination room you must give this book to the invigilator. If you do not, you may lose all the marks for this paper.

SCOTTISH QUALIFICATIONS AUTHORITY

LI 0300/402 6/27570

DO NOT WRITE IN THIS MARGIN

Marks | KU | PS

1. The graph shows the growth curve of a population of bacteria in a fermenter at 30 °C over a 24 hour period.

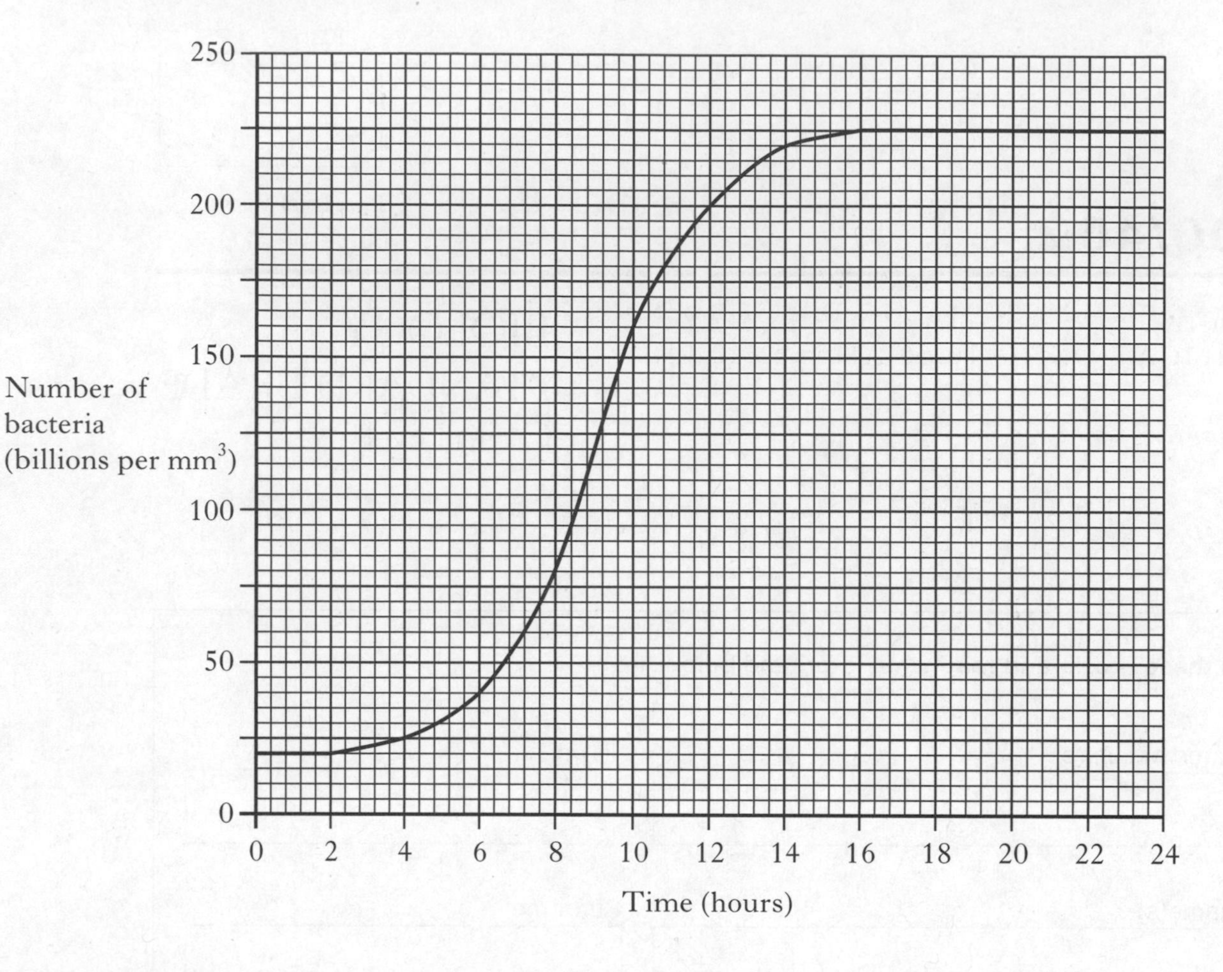

(*a*) (i) How long did it take for the population to double its starting size?

__________ hours **1**

(ii) In which two-hour period was there the greatest increase in the number of bacteria?

Between __________ hours and __________ hours **1**

(iii) Between which times did the rate of production of new bacteria exceed the death rate?

Between __________ hours and __________ hours **1**

(*b*) (i) Describe the relationship between the number of bacteria in the population and time.

____

____ **2**

DO NOT WRITE IN THIS MARGIN

Marks KU PS

1. (*b*) (continued)

(ii) Explain the changes in the shape of the growth curve between 10 hours and 18 hours in terms of the factors that affect population growth.

____

______________________________ **2**

(iii) Draw a second line on the graph to predict the growth in population if the fermenter had been kept at a temperature of 20 °C.

(An additional graph, if needed, will be found on page 22.) **1**

(*c*) (i) Before setting up the fermenter, all the apparatus was heated to 150 °C for 15 minutes to eliminate any contamination by resistant cells of bacteria and fungi.

What name is given to these resistant cells?

______________________________ **1**

(ii) The fermenter was stirred before removing the samples used to estimate the numbers of bacteria. How would this minimise possible errors in the results?

____

______________________________ **1**

[Turn over

DO NOT WRITE IN THIS MARGIN

Marks | KU | PS

2. (*a*) The table shows the percentage germination of four crop plants over a range of temperatures.

Temperature (°C)	*Percentage germination of crop plants*			
	Carrots	*Cauliflower*	*Okra*	*Spinach*
0	0	0	0	83
5	48	0	0	96
10	93	58	0	91
15	95	60	74	80
20	96	65	89	52
25	95	53	93	28
30	90	45	88	14
35	74	0	85	0
40	0	0	35	0

(i) Which **two** crop plants are able to germinate over the widest range of temperatures?

1 ____________________ 2 ____________________ **1**

(ii) Complete the table below by adding the correct heading, units and values to show the optimum germination temperature for each of the crop plants.

Crop plant	

2

(iii) Suggest which crop plant would germinate best in a hot climate.

____________________ **1**

DO NOT WRITE IN THIS MARGIN

Marks KU PS

2. (*a*) **(continued)**

(iv) What is the minimum number of spinach seeds which should be sown at 15 °C in order to produce 1000 seedling plants?

Space for calculation

Number of seeds ________ 1

(v) On the grid below, complete a line graph of the change in percentage germination of cauliflower seeds with temperature.

(An additional graph, if needed, will be found on page 22.)

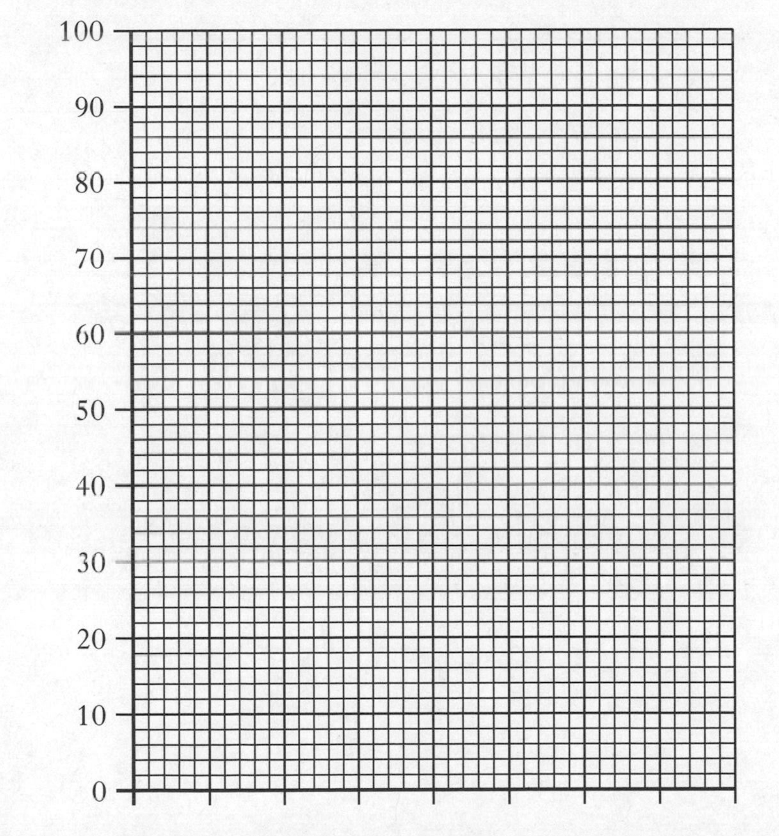

2

(*b*) Which part of a seed starts to develop during germination?

________________________ 1

(*c*) Describe the changes in percentage germination of seeds that occur over a range of temperatures.

____

________________________ 2

DO NOT WRITE IN THIS MARGIN

Marks	KU	PS

3. (*a*) Sexual reproduction in flowering plants depends on the processes of pollination and fertilisation.

Describe the events from the time a pollen grain of the correct species lands on the stigma, until fertilisation takes place in the ovary.

____

____

______________________________ **2**

(*b*) Plant growers can propagate plants by artificial methods such as cuttings and grafting.

Give **two** advantages to the plant growers of artificial propagation of flowering plants.

1 ______________________________

2 ______________________________ **2**

DO NOT WRITE IN THIS MARGIN

Marks KU PS

4. (*a*) Fertilisation is the fusion of gametes and can be either internal or external in animals.

Explain why it is necessary for some animals to use internal fertilisation.

____

____ **1**

(*b*) A human fetus develops inside the mother's uterus, attached to the placenta.

Name **one** substance which passes across the placenta from mother to fetus.

________________________________ **1**

(*c*) Some animal species take more care of their young than others.

Describe the relationship between the degree of parental care and the number of eggs that are produced at any one time by different species.

____

____ **1**

[Turn over

DO NOT WRITE IN THIS MARGIN

Marks KU PS

5. (*a*) Food is moved along the alimentary canal by the action of circular muscles.

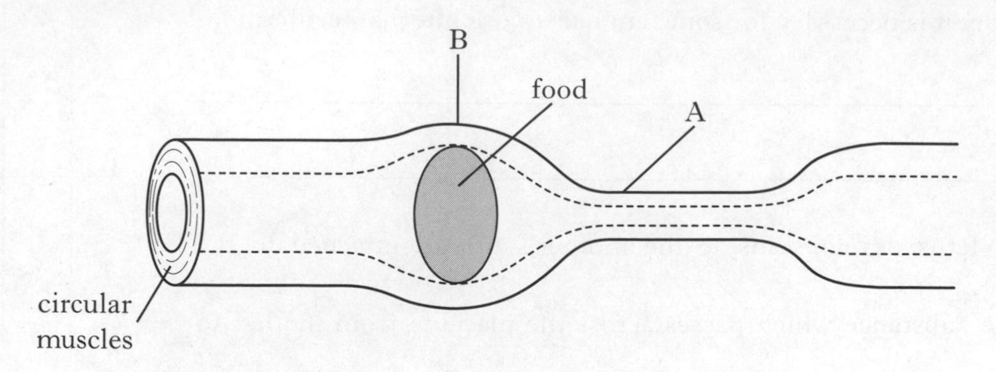

(i) What name is given to this movement of food?

_______________ **1**

(ii) Describe the state of the muscles at positions A and B in the diagram.

A _______________

B _______________ **1**

(*b*) When food reaches the stomach it is mixed with digestive juices.
Name **one** other organ that produces digestive juices.

_______________ **1**

(*c*) The table shows some of the daily vitamin and mineral requirements of teenagers.

	Daily requirement (mg)				
Sex	*Vitamin B3*	*Vitamin C*	*Calcium*	*Iron*	*Zinc*
Girls	13	40	1000	15	7
Boys	17	40	800	11	10

(i) Which substance is required in equal quantities by both sexes?

_______________ **1**

(ii) Which substances are required in greater quantities by boys?

_______________ **1**

DO NOT WRITE IN THIS MARGIN

Marks	KU	PS

5. (c) (continued)

(iii) Calculate the daily requirement of calcium for girls compared to boys as a simple whole number ratio.

Space for calculation

_____ : _____ 1

girls : boys

(d) The graph shows the changes in the vitamin C content of potatoes during storage.

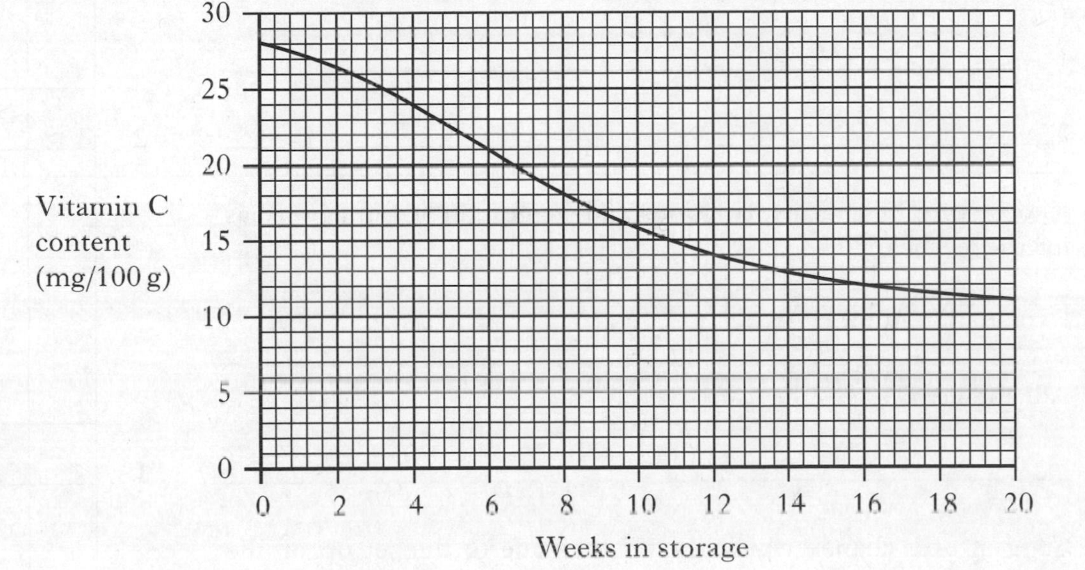

(i) How long did it take for the vitamin C content to fall to half of the original value?

__________ weeks 1

(ii) After six weeks in storage, what percentage of the original vitamin C still remains in the potato?

Space for calculation

__________ % 1

[Turn over

DO NOT WRITE IN THIS MARGIN

Marks KU PS

6. (*a*) In an investigation into behaviour, five leeches were placed in water in a shallow rectangular dish as shown in the diagram.

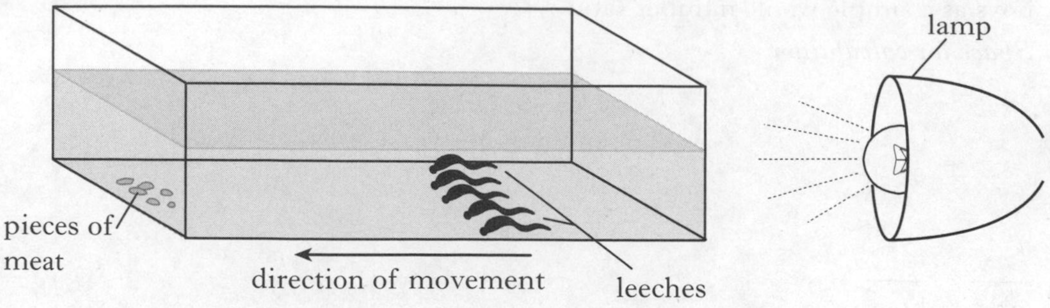

lamp

pieces of meat

direction of movement

leeches

(i) During the investigation the leeches moved in the direction shown.

Give **two** possible explanations for this response.

1 ____

2 ____ **2**

(ii) Choose **one** of your explanations and suggest an advantage it has for the leeches.

Explanation number ________

Advantage ____

____ **1**

(iii) Suggest **one** change which should be made to the set up of the investigation so that only one valid conclusion could be drawn from the leeches' response, assuming the direction of movement stays the same.

____

____ **1**

(*b*) (i) Swallows migrate from Britain to Africa in the autumn.

Explain how this behaviour benefits the swallows.

____

____ **1**

(ii) Migration is an example of a type of behaviour that is repeated regularly. What name is given to this type of behaviour?

__________________________ **1**

DO NOT WRITE IN THIS MARGIN

Marks KU PS

7. In an investigation, three 25 g samples of sultanas were put into separate beakers of distilled water, as shown below.

distilled water

sultanas

After 24 hours, the sultanas were removed from the water, blotted on filter paper and reweighed. The results are shown in the table.

Sample	*Mass after 24 hours* (g)	*Percentage change in mass*
1	32·5	30·0
2	32·2	28·8
3	32·4	

(*a*) Complete the table with the percentage change in mass of the sultanas in sample 3.

Space for calculation

1

(*b*) The change in mass of the sultanas was caused by the movement of water.

(i) Name this process.

______________________________ 1

(ii) Explain the results in terms of water concentrations.

____

______________________________ 1

(*c*) Which of the following is the best reason for blotting the sultanas before reweighing?

Tick the correct box.

☐ To stop them sticking together ☐ To remove external sugar solution

☐ To remove external water ☐ To make sure the sultanas were dried 1

DO NOT WRITE IN THIS MARGIN

Marks | KU | PS

8. The following statements refer to stages in mitosis.

1 Chromosomes become visible as pairs of chromatids.

2 Spindle fibres form.

3 ____

____

4 Chromatids are pulled to opposite ends of the cell.

5 The nuclear membranes form.

6 The cytoplasm divides and two daughter cells are formed.

(*a*) Complete the sequence by writing in a description of the missing stage. **1**

(*b*) After mitosis, the daughter cells have the same number of chromosomes as the parent cell.
Explain why this is important.

____

____ **1**

DO NOT WRITE IN THIS MARGIN

Marks | KU | PS

9. Read the following passage and answer the questions using information from it.

Adapted from *The Herald*, October 2003

Scientists say that the North Sea is becoming too hot for many of the fish which are included in the normal Scottish diet. Experts are blaming global warming for driving the plankton, on which the fish depend, into more northern waters. As a result, stocks of cod and salmon are in danger of collapse. At the same time, more exotic species such as red mullet, horse mackerel and black bream are increasing off the east coast of Britain.

Sand eels are also dwindling in number, and this may be having a knock-on effect on the coastal birds which feed on them. A survey of their habitats showed breeding rates for puffins, kittiwakes, guillemots and razorbills to be the lowest on record.

These trends are based on the monitoring of plankton populations. They may help to explain why a reduction in fishing has not led to a full recovery of fish populations.

Two particular episodes are blamed. The first occurred in the late 1970s and was caused by an inflow of low-temperature, low-salinity water from the North Atlantic. This was due to a high release of Arctic ice into the ocean. The second occurred in the 1980s, and this time it was an inflow of water at higher temperatures and high salinity.

(*a*) What effect is global warming having on the plankton in the North Sea?

______________________________ **1**

(*b*) Name **two** fish species which are decreasing in numbers in the North Sea.

1 ________________ 2 ________________ **1**

(*c*) Suggest a reason why exotic fish species are increasing in number off the east coast of Britain.

____

______________________________ **1**

(*d*) Explain the possible link between global warming and the expected reduction in the numbers of coastal birds.

____

______________________________ **1**

(*e*) In what **two** ways did the water which caused problems in the 1980s differ from that which caused problems in the 1970s?

1 ______________________________

2 ______________________________ **1**

DO NOT WRITE IN THIS MARGIN

Marks KU PS

10. The diagram represents the structures involved in a reflex action which occurs when a finger touches a flame.

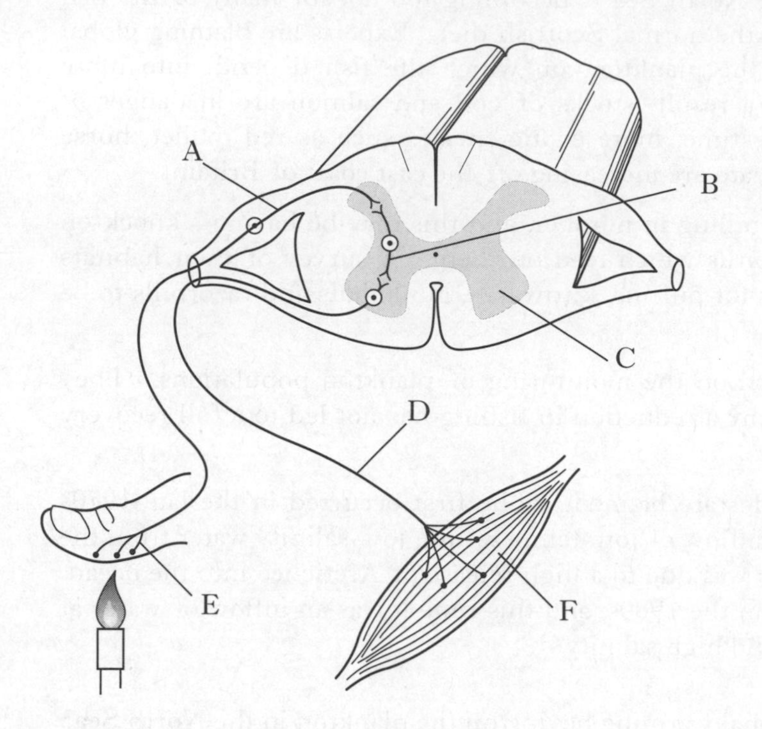

(*a*) Complete the table below with the correct letters from the diagram to identify the stages of the reflex action and with a description of the missing stage.

Stage	*Letter*
Stimulus detected by sensory receptor	
Information sent along a sensory nerve cell	
	B
Impulse sent along motor nerve cell	
Response made by effector organ	

2

DO NOT WRITE IN THIS MARGIN

Marks KU PS

10. (continued)

(*b*) An investigation was carried out on the response of the pupil of the eye. A volunteer was seated in a dark room and a torch was switched on. The diameter of the volunteer's pupil was measured.

This was repeated at different distances from the volunteer.

The results are shown on the graph below.

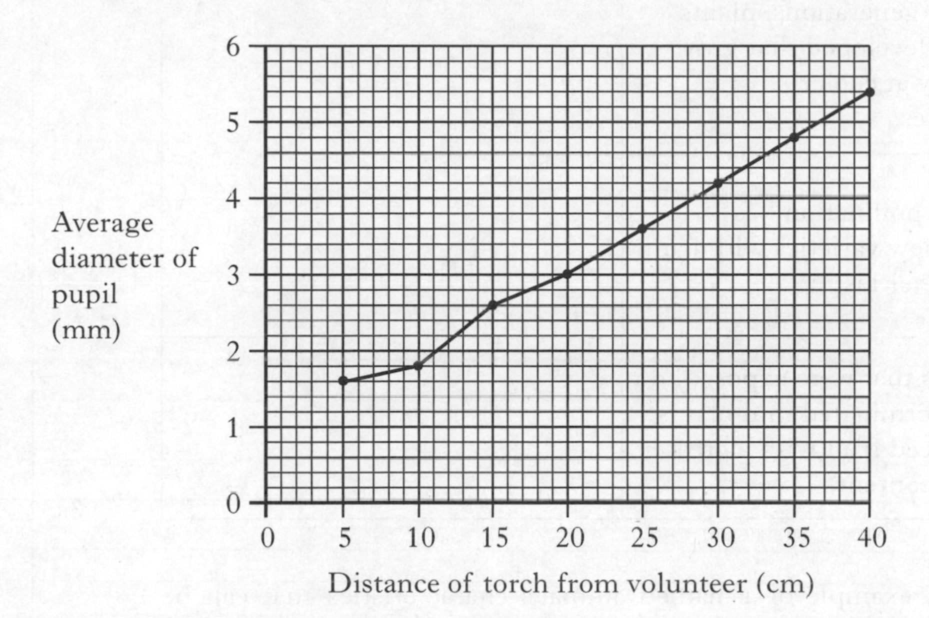

(i) What was the diameter of the volunteer's pupil when the torch was 15 cm away from the eye?

_______ mm **1**

(ii) Draw **one** conclusion from these results.

____

____ **1**

(iii) The investigation was repeated several times and the average values of the pupil diameters were calculated before the graph was drawn.

Why is this good experimental procedure?

____

____ **1**

[Turn over

DO NOT WRITE IN THIS MARGIN

Marks KU PS

11. (*a*) The following table gives examples of improvements in tomato plants.

Complete the table to show whether each improvement is a result only of mutation or if it also involves selective breeding.

Improvement	*Only mutation/ Involves selective breeding*
Over many generations, plants have been developed that grow successfully at cooler temperatures.	
Controlled pollination has produced new varieties with fruit that is sweeter tasting.	
Some seeds that were exposed to radiation germinated into plants that produced fruit with a greater vitamin C content.	

2

(*b*) Describe an example of a named animal's characteristics that can be improved by selective breeding.

Animal ______________________________

Description of improved characteristic ______________________________

______________________________ 1

DO NOT WRITE IN THIS MARGIN

Marks KU PS

12. Tay-Sachs disease is an inherited condition which affects the nerves. Different forms of the same gene determine its effect.

T (dominant) represents the normal form of the gene.
t (recessive) represents the form of the gene which causes the disease.

The family tree diagram shows a pattern of inheritance of the disease.

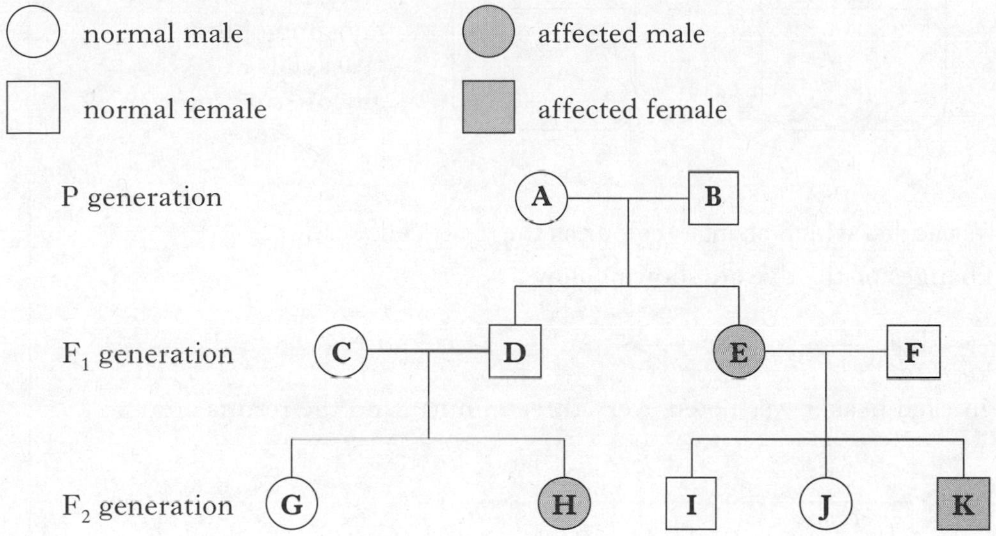

normal male — affected male
normal female — affected female

P generation: A, B
F_1 generation: C, D, E, F
F_2 generation: G, H, I, J, K

(*a*) (i) Complete the table by writing the genotypes of persons **A**, **D** and **K**.

Person	*Genotype*
A	
D	
K	

2

(ii) A carrier of the disease is someone who does not show the symptoms of the disease but can pass it to their offspring.

Give the letter of **one** person from the F_2 generation who must be a carrier of the disease.

Letter ____________ 1

(iii) What kind of variation is shown by Tay-Sachs disease? Explain your answer.

Variation ____________

Explanation ____________

____________ 1

(*b*) What name is given to the different forms of the same gene?

____________ 1

DO NOT WRITE IN THIS MARGIN

Marks KU PS

13. The investigation below was used to compare the respiration rates of immobilised and non-immobilised yeast cells.

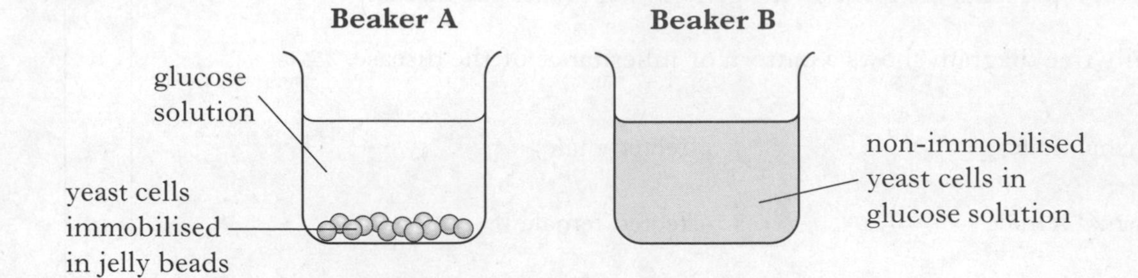

A blue dye was added which changes colour as the yeast cells respire.
The colour changes of the dye are shown below.

blue ⟶ lilac ⟶ mauve ⟶ pink ⟶ colourless

The colour in each beaker was noted every three minutes and the results are shown below.

Time (minutes)	*Beaker A*	*Beaker B*
0	blue	blue
3	blue	lilac
6	lilac	mauve
9	lilac	mauve
12	mauve	colourless
15	mauve	colourless
18	pink	colourless
21	colourless	colourless

(*a*) (i) In which beaker did the yeast cells respire faster?
Give a reason for your answer.

Beaker ____________

Reason ____ **1**

(ii) Suggest a time when the dye in beaker B might have been pink.

____________ minutes **1**

DO NOT WRITE IN THIS MARGIN

Marks KU PS

13. (*a*) (continued)

(iii) Give **two** precautions that would have to be taken to ensure a valid comparison could be made between the two beakers.

1 ______________________________

2 ______________________________ **2**

(*b*) Immobilised cells are used in some industrial processes.

Describe **one** advantage of using immobilised cells.

____

______________________________ **1**

(*c*) The table gives information about respiration in yeast.

Tick the boxes to show whether each statement refers to aerobic respiration, anaerobic respiration or both.

Statement	*Aerobic*	*Anaerobic*
Oxygen is used up.		
Alcohol is produced.		
Maximum energy is released.		
Carbon dioxide is produced.		

2

[Turn over

DO NOT WRITE IN THIS MARGIN

Marks KU PS

14. Cellulase is an enzyme which is produced by some soil micro-organisms. It breaks down cellulose into simple sugars. Cellulose is present in plant cell walls.

10 cm^3 samples of cellulose paste were mixed with three different liquids and left for 24 hours. The time taken for 5 cm^3 of each cellulose mixture to run through a syringe was recorded. The results are shown in the table.

Sample	*Liquid added to cellulose paste*	*Time for* 5 cm^3 *to run through* (seconds)
A	1 cm^3 cellulase solution	126
B	1 cm^3 water	375
C	1 cm^3 soil water	200

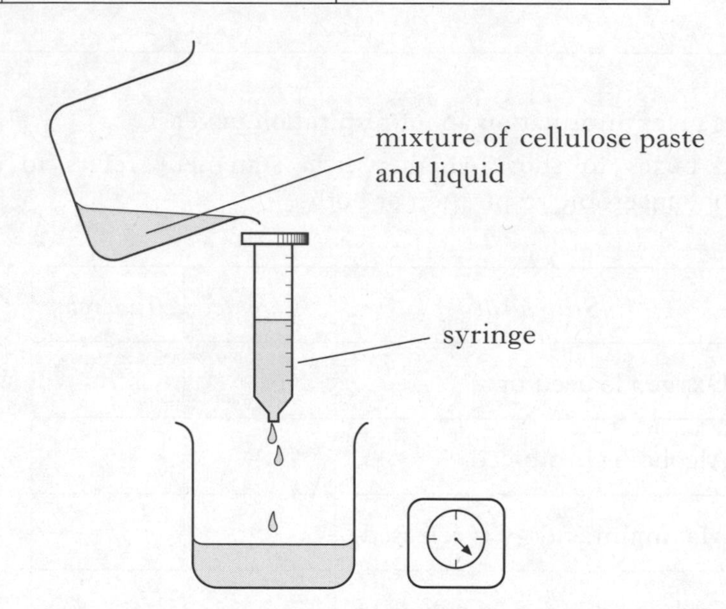

(*a*) (i) Using the results from samples A and B, describe the effect of cellulase on the thickness of cellulose paste.

______________________________________ **1**

(ii) Using the results, what can be concluded about soil water?

______________________________________ **1**

(*b*) (i) The samples were left in a warm place to provide optimum conditions for the enzyme.

Explain what is meant by the term *optimum conditions*.

______________________________________ **1**

(ii) Cellulase enzyme is specific for cellulose.

Explain what is meant by the term *specific*.

______________________________________ **1**

DO NOT WRITE IN THIS MARGIN

Marks KU PS

15. (*a*) The following chart shows the volume of air present in the lungs of a person during a period of normal breathing.

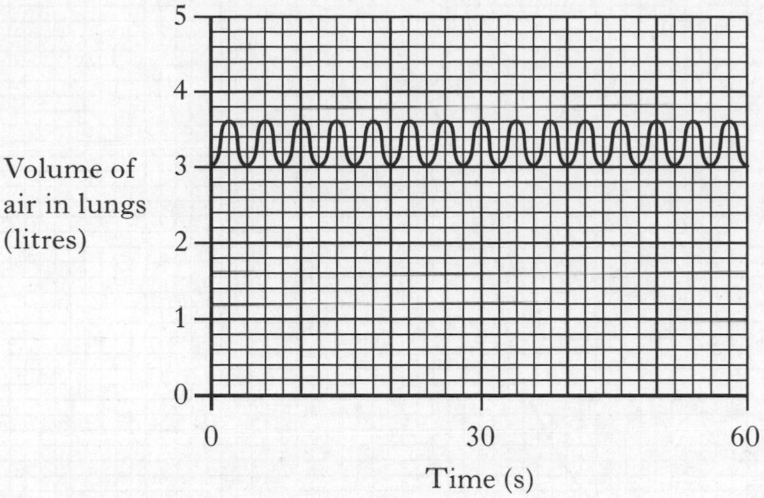

Volume of air in lungs (litres)

Time (s)

(i) What is the volume of air inhaled in one breath?

__________ litres **1**

(ii) What is the person's breathing rate?

__________ breaths per minute **1**

(*b*) (i) Regular exercise improves the efficiency of the lungs.

What other body system, essential for muscle activity, also benefits from regular exercise?

____ **1**

(ii) Explain why increased efficiency of the lungs results in an improved recovery time following exercise.

____

____ **1**

[*END OF QUESTION PAPER*]

[Turn over

ADDITIONAL GRAPH PAPER FOR QUESTION 1(*b*)(iii)

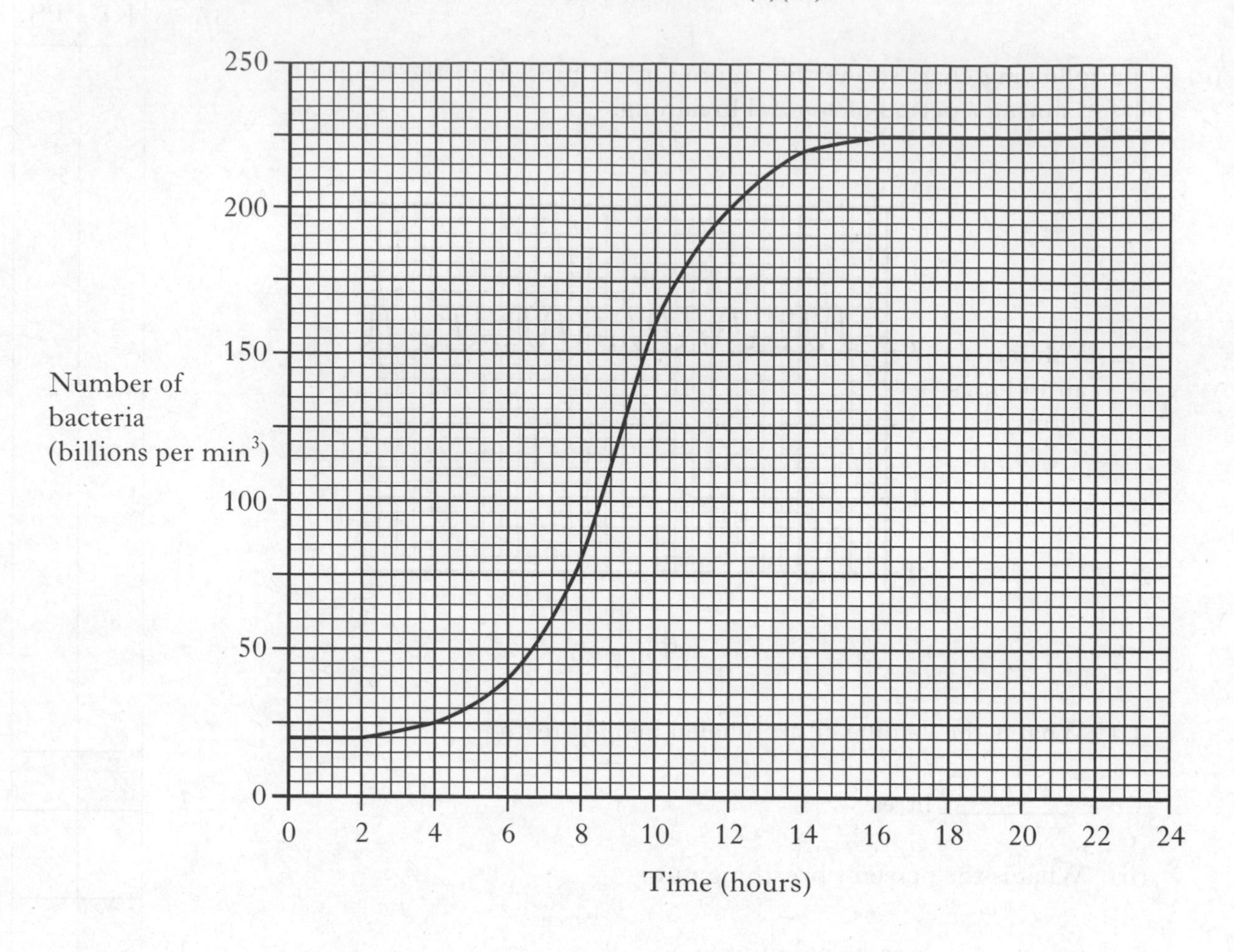

ADDITIONAL GRAPH PAPER FOR QUESTION 2(*a*)(v)

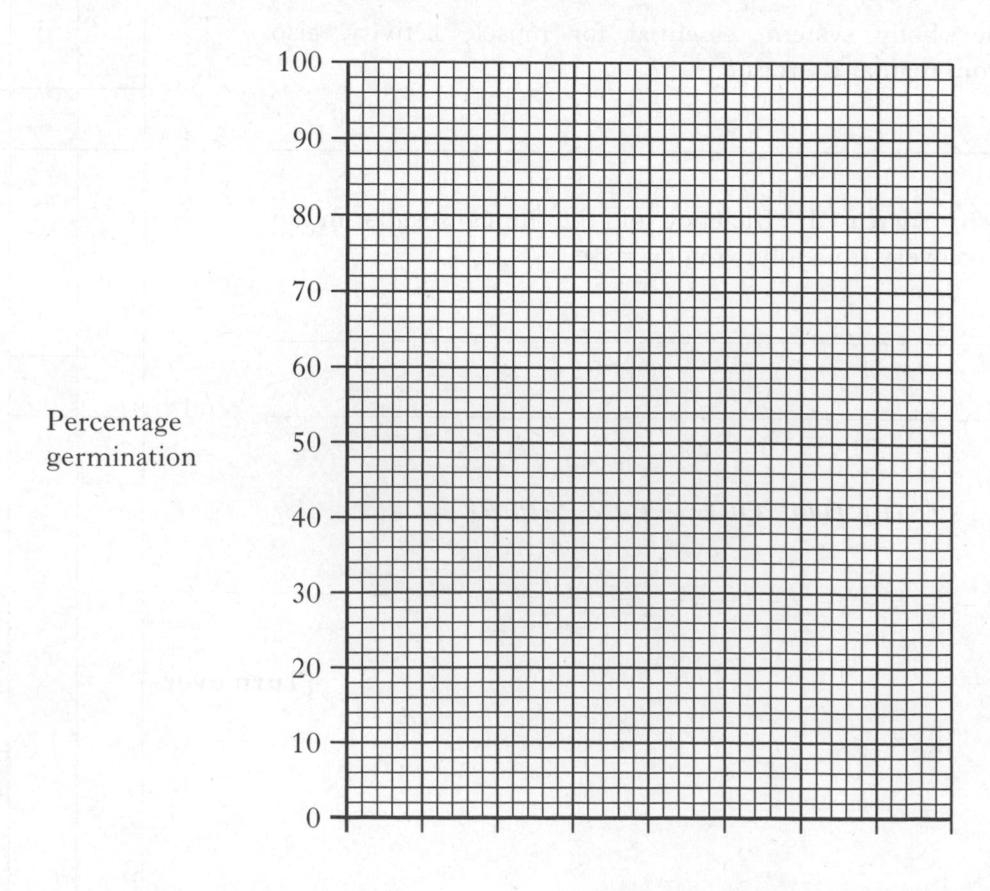

STANDARD GRADE | CREDIT

2007

[BLANK PAGE]

FOR OFFICIAL USE

C

KU	PS

Total Marks

0300/402

NATIONAL QUALIFICATIONS 2007

MONDAY, 21 MAY
10.50 AM – 12.20 PM

BIOLOGY
STANDARD GRADE
Credit Level

Fill in these boxes and read what is printed below.

Full name of centre

Town

Forename(s)

Surname

Date of birth
Day Month Year

Scottish candidate number

Number of seat

1 All questions should be attempted.

2 The questions may be answered in any order but all answers are to be written in the spaces provided in this answer book, and must be written clearly and legibly in ink.

3 Rough work, if any should be necessary, as well as the fair copy, is to be written in this book. Additional spaces for answers and for rough work will be found at the end of the book. Rough work should be scored through when the fair copy has been written.

4 Before leaving the examination room you must give this book to the invigilator. If you do not, you may lose all the marks for this paper.

SCOTTISH QUALIFICATIONS AUTHORITY

DO NOT WRITE IN THIS MARGIN

Marks KU PS

1. The graph shows the changes in the population of bacteria in a fermenter.

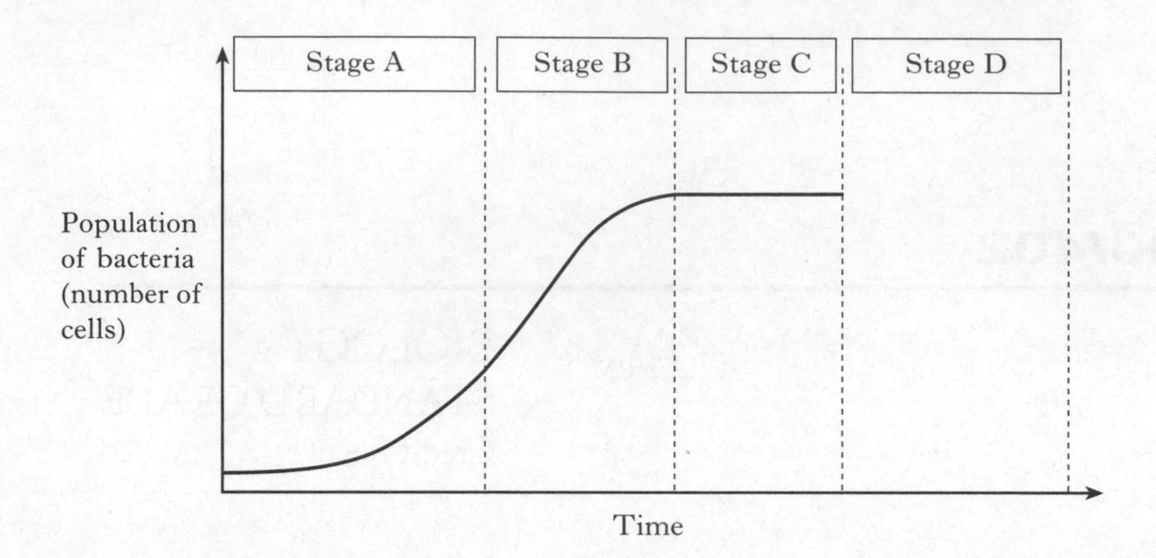

(*a*) (i) Describe the changes in population of the bacteria during Stage B.

____

____ **1**

(ii) Give a reason for the changes in population shown during Stage B on the graph.

____

____ **1**

(iii) Complete Stage D on the graph to show the effect of an increasing death rate on the population of bacteria. **1**

(*b*) Some bacteria can be grown on industrial waste materials to provide valuable products, such as animal foodstuffs.

State **one** way in which the nutritional value of the product has been increased.

____ **1**

DO NOT WRITE IN THIS MARGIN

Marks KU PS

2. The diagram shows some of the stages in the nitrogen cycle.

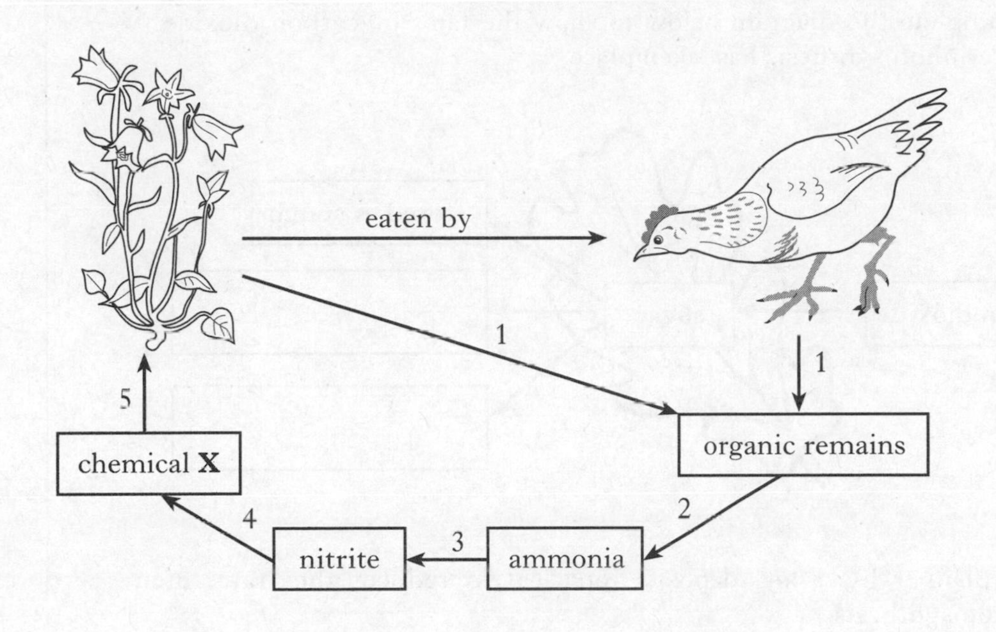

(*a*) Complete the table by giving a number from the diagram to match each of the named stages.

Stage	*Number*
Absorption	
Death	
Nitrification	
Decomposition	

2

(*b*) Name chemical **X**.

________________________________ **1**

(*c*) Name the type of organism responsible for Stage 3.

________________________________ **1**

[Turn over

DO NOT WRITE IN THIS MARGIN

Marks KU PS

3. (*a*) Carbon dioxide is used during photosynthesis to produce sugar.

(i) Complete the diagram below to show the fates of carbon dioxide after photosynthesis has taken place.

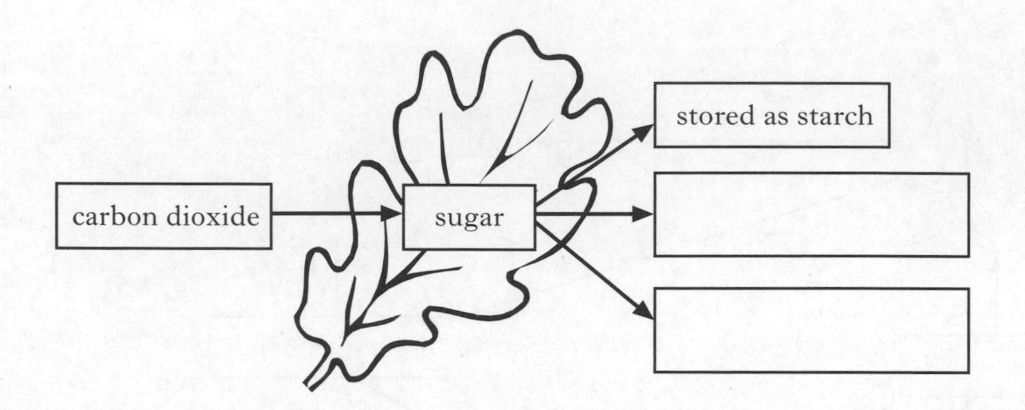

2

(ii) Explain why soot deposits on leaves reduce the rate of photosynthesis.

____

____ **1**

(*b*) (i) Draw an **X** on the following diagram to show where the pollen tube reaches when its growth is completed.

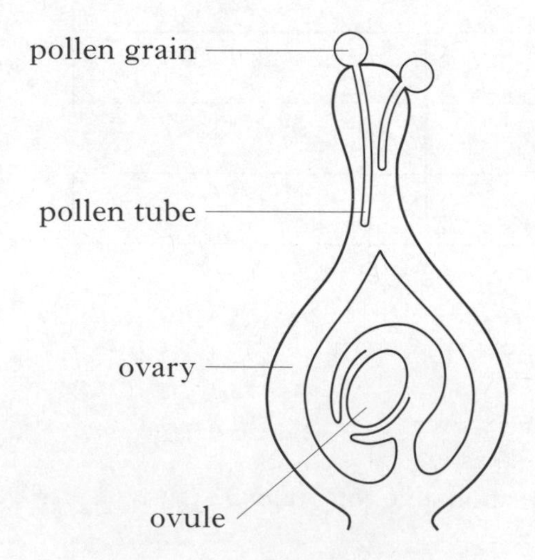

1

(ii) Describe the function of the pollen tube.

____

____ **1**

DO NOT WRITE IN THIS MARGIN

Marks KU PS

3. (continued)

(*c*) Tropical rain forests are being destroyed to clear land for farming. This leads to a reduction in the number of plant species.

Explain why this might lead to the extinction of some animal species.

____ **1**

(*d*) The diagrams show features of some newly discovered plants.

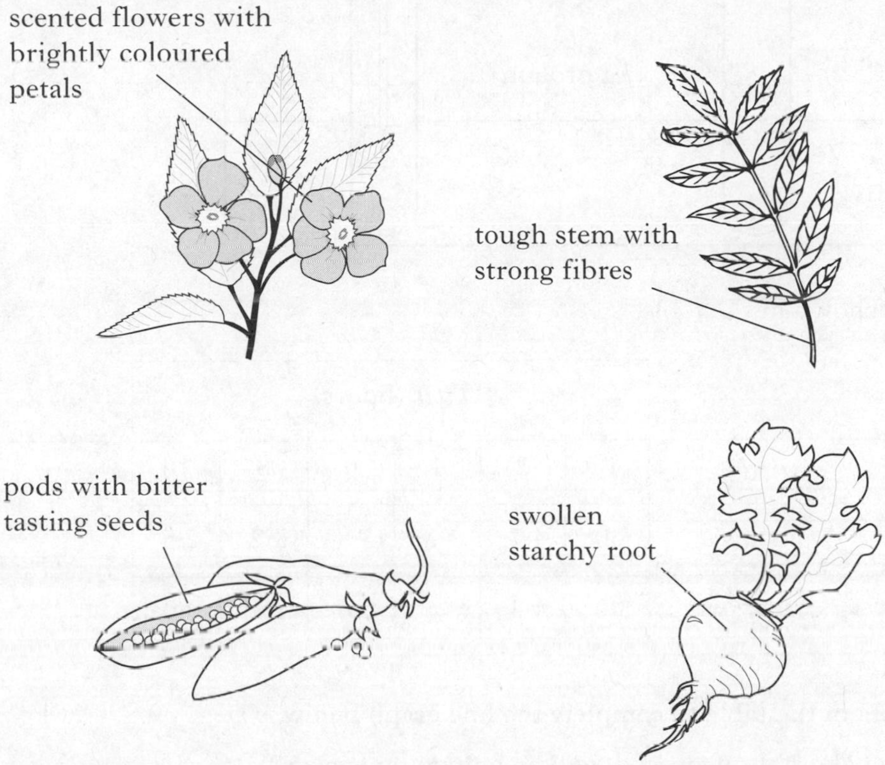

scented flowers with brightly coloured petals

tough stem with strong fibres

pods with bitter tasting seeds

swollen starchy root

Select **one** of the plant features and describe a likely use for it.

Plant feature ____________________________________

Likely use _____________________________________

____ **1**

[Turn over

DO NOT WRITE IN THIS MARGIN

Marks KU PS

4. The following investigation was set up to examine the effects of stirring on the digestion of protein.

Each piece of protein was weighed every two hours.

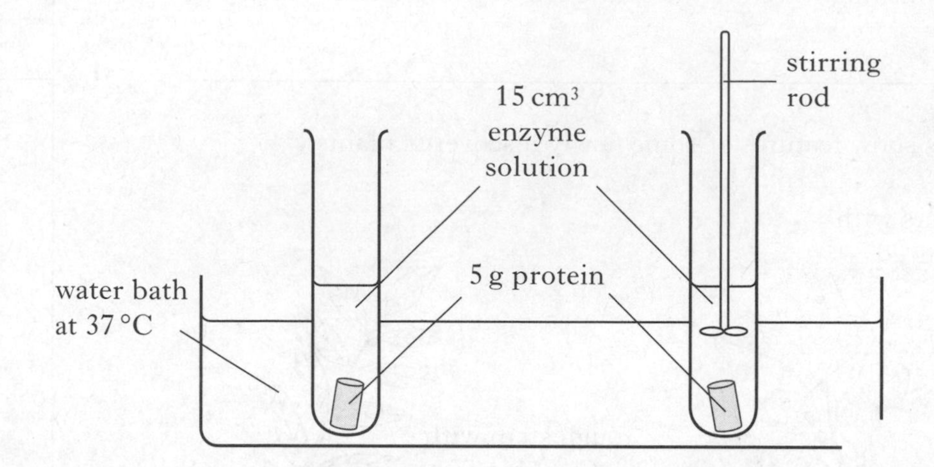

The results are shown in the table.

		Time (hours)					
		0	2	4	6	8	10
Mass of protein (g)	*not stirred*	5·0	4·7	4·3	3·8	3·2	2·5
	stirred	5·0	4·4	3·6	2·6	1·4	0·0

(*a*) Use the data in the table to complete the line graph below.

(An additional graph, if needed, will be found on page 25.)

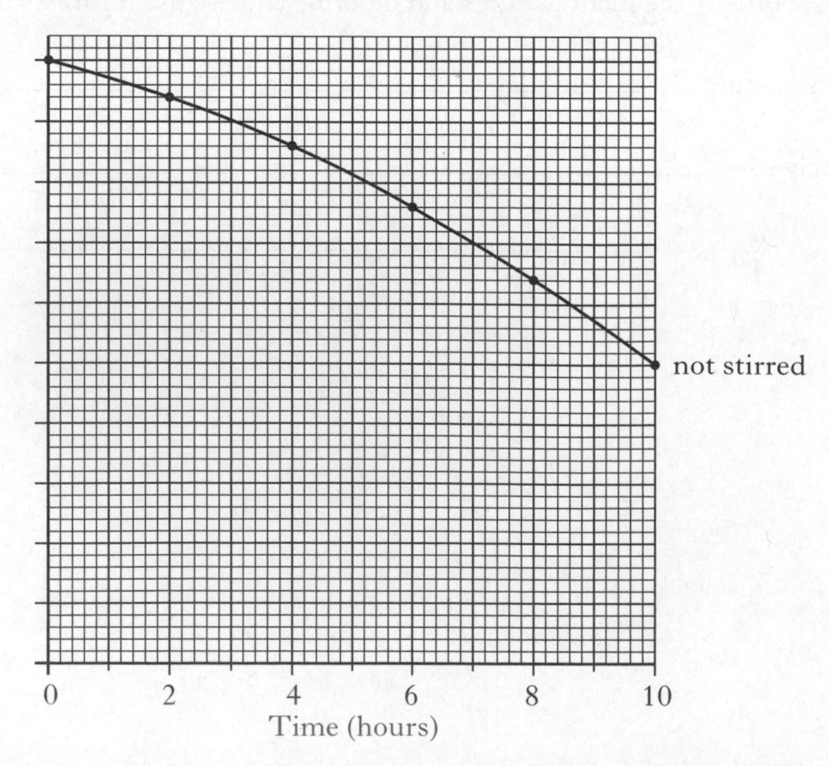

2

DO NOT WRITE IN THIS MARGIN

Marks KU PS

4. (continued)

(*b*) Which type of enzyme would produce the results shown?

______________________ **1**

(*c*) When the protein was completely digested, no solid material remained in the tube. Explain why.

______________________ **1**

(*d*) Name **one** factor, not already mentioned, which would need to be the same in each tube at the start of the investigation.

______________________ **1**

(*e*) Suggest how the investigation could be improved to provide a more reliable measurement of the difference which stirring made.

______________________ **1**

(*f*) Stirring increased the rate at which the protein was digested. Explain why this happened.

______________________ **1**

(*g*) In the body, the stomach achieves a similar effect to stirring. Describe how this happens.

______________________ **1**

[Turn over

DO NOT WRITE IN THIS MARGIN

Marks | KU | PS

5. The diagram represents a microscopic part of a kidney.

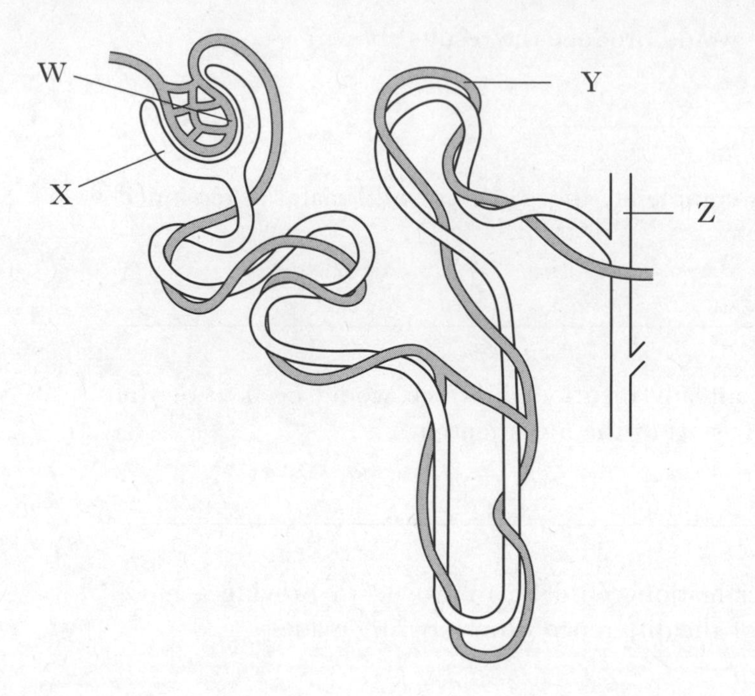

(*a*) Complete the table to show the names and functions of the structures shown on the diagram.

	Name	*Function*
W	glomerulus	
X		collection of filtrate
Y		reabsorption
Z	collecting duct	

2

DO NOT WRITE IN THIS MARGIN

5. (continued)

Marks KU PS

(*b*) The table shows information about kidney function.

Fluid	*Component* (g per 100cm^3)				
	urea	*glucose*	*amino acids*	*salts*	*proteins*
blood plasma	0·03	0·10	0·05	0·9	8·0
glomerular filtrate	0·03	0·10	0·05	0·9	none
urine	1·75	none	none	0·90–3·60	none

(i) In which organ is urea produced and how is it transported to the kidneys?

Organ ________________

Means of transport ________________ 1

(ii) Name **one** component in the table which can pass through the wall of the glomerulus, and **one** component which cannot.

Can pass through ________________

Cannot pass through ________________ 1

(*c*) In one investigation, the kidneys of an adult male were found to filter 1254 cm^3 of blood per minute. This produced 114 cm^3 of filtrate per minute and 1·2 cm^3 of urine per minute.

(i) Express these volumes as a simple whole number ratio.

Space for calculation

_______ : _______ : _______
blood filtrate urine 1

(ii) Using the results of this investigation and information from the table, calculate the mass of urea which would be excreted by this person in 24 hours.

Space for calculation

__________ g 1

[Turn over

DO NOT WRITE IN THIS MARGIN

Marks KU PS

6. The brown shrimp is found all round our coastline.

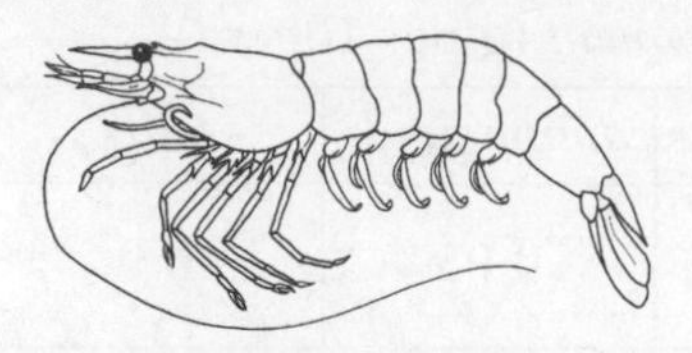

The graph shows shrimp activity and changes in their environment over a 48 hour period.

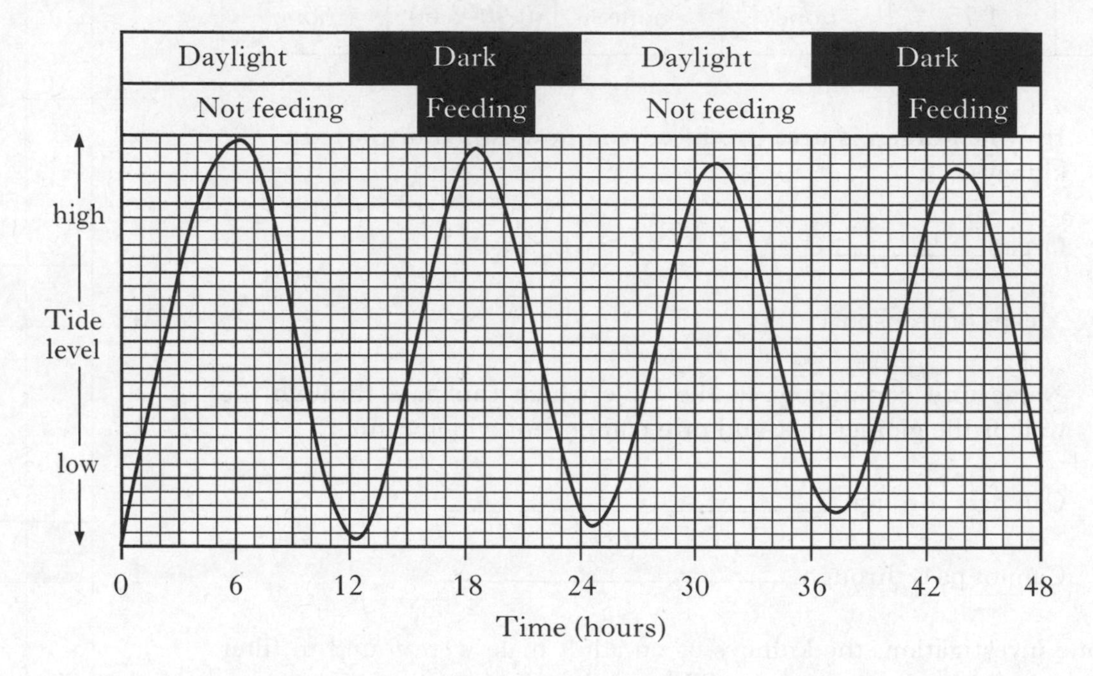

(*a*) How many high tides occurred during the two days shown?

____________ **1**

(*b*) Describe the conditions necessary for the shrimps to feed.

____

____ **2**

(*c*) Explain the significance of the behaviour shown to the survival of the shrimps.

____

____ **1**

DO NOT WRITE IN THIS MARGIN

Marks KU PS

7. A flower petal was examined under the microscope and then placed in a concentrated salt solution for 30 minutes. It was then re-examined under the microscope.

The diagrams show a cell from the petal before and after being in the solution.

before after

(*a*) (i) The movement of water caused the change in the appearance of the cell. What name is given to this movement of water?

________________________ **1**

(ii) In terms of water concentration, explain **why** this movement of water took place.

____

____ **1**

(*b*) Name **one** substance, other than water, which must be able to pass into a cell for its survival.

________________________ **1**

(*c*) The diagram below shows a group of cells as seen under a microscope. The field of view was 2 mm in diameter.

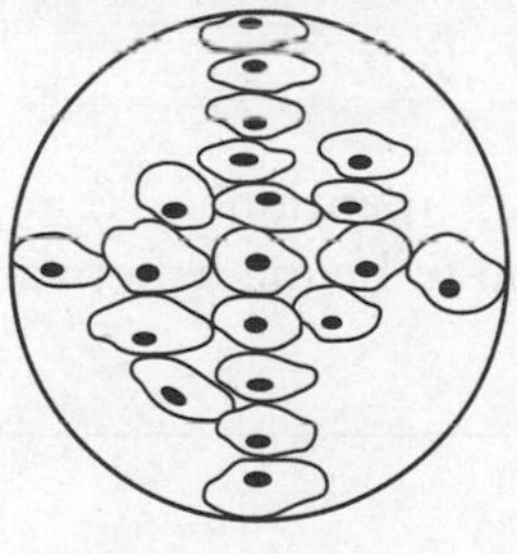

Calculate the average length and width of the cells.

Space for calculation

Average length ______ mm

Average width ______ mm **1**

[Turn over

DO NOT WRITE IN THIS MARGIN

Marks | KU | PS

8. (*a*) The diagram shows a method used to investigate the energy content of a variety of foods.

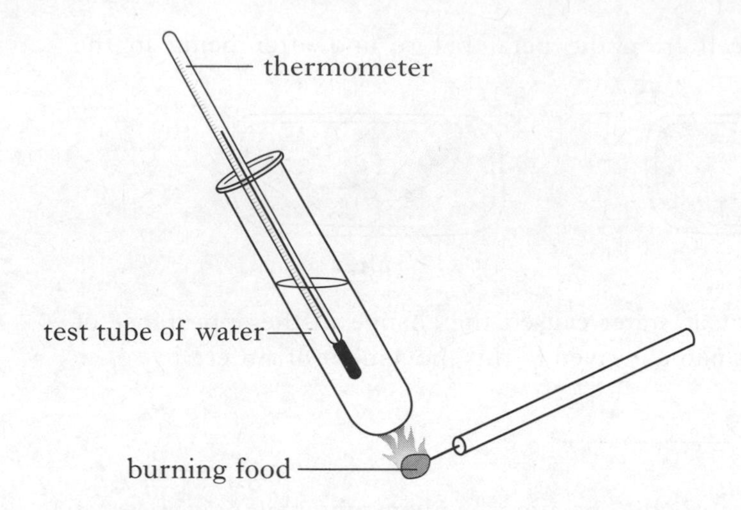

The rise in temperature can be used to calculate the energy content of each food in kilojoules.

The results are shown in the table.

Type of food	*mass* (g)	*energy content* (kilojoules)
cheese	1·0	17·0
fish	1·0	0·5
steak	1·0	13·9
carrot	1·0	1·8
apple	1·0	2·5

(i) State **two** factors, not already mentioned, that should be kept constant for a valid comparison to be made between the foods.

1 ____________________

2 ____________________ **2**

(ii) Suggest why the energy contents found in the investigation might not have been as high as expected.

____

____________________ **1**

DO NOT WRITE IN THIS MARGIN

Marks | KU | PS

8. (*a*) (continued)

(iii) The energy content of each food was calculated using the following formula.

Energy content (kilojoules) = temperature rise × 0·21

Calculate the energy content of 1g of chicken, if it raised the temperature of the water by 30 °C.

Space for calculation

_______ kilojoules per gram **1**

(*b*) Give **one** reason, other than providing heat, why cells need energy from food.

____ **1**

(*c*) Which component of food provides most energy per gram?

______________________________ **1**

[Turn over

DO NOT WRITE IN THIS MARGIN

Marks KU PS

9. The diagram below shows a cross-section through a joint.

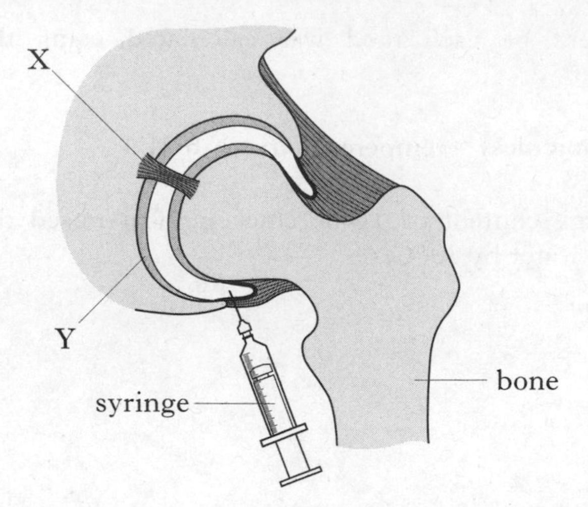

(*a*) Name and describe the functions of parts X and Y on the diagram.

Part X Name ____________

Function ______________________________________

______________________________________ **1**

Part Y Name ____________

Function ______________________________________

______________________________________ **1**

(*b*) Some of the synovial fluid from inside a joint can be removed for medical tests using a syringe as shown in the diagram.

(i) Name the part of the joint which produces the synovial fluid and describe the function of the fluid.

Produced by ______________________________________

Function ______________________________________ **1**

DO NOT WRITE IN THIS MARGIN

Marks KU PS

9. (*b*) (continued)

(ii) The table below describes the features of the fluid which lead to the diagnosis of several joint abnormalities.

		Feature of synovial fluid		
		Viscosity	*Cloudiness*	*Colour*
Diagnosis	*Normal*	high	zero	light yellow
	Inflammation	low	slight	dark yellow
	Infection	low	high	dark yellow
	Blood leakage	intermediate	high	pink

Use the information from the table to complete the paired statement key to identify the diagnoses.

1. Fluid pink .... Blood leakage

Fluid not pink .... go to 2

2. Low viscosity .... [ ]

High viscosity .... [ ]

3. [ ] .... Infection

[ ] .... [ ]

2

[Turn over

DO NOT WRITE IN THIS MARGIN

Marks | KU | PS

10. Read the following passage and answer the questions based on it.

Invasion of the Chinese Mitten Crab

Adapted from *Biological Sciences Review*, Volume 15, Number 2.

The Chinese mitten crab, *Eriocheir sinensis*, lives in fresh water as an adult, but it breeds in the lower reaches of estuaries and spends part of its early life in seawater.

It looks different from other crabs. Its claws are covered in a coating of fine brown hairs resembling mittens. This type of crab is a problem because it burrows into river banks, causing them to collapse and silt up river channels.

The mitten crab is not native to Europe. They were recorded in the River Thames in the 1930s. Their larvae may have been transported to the river in ships' ballast water and released during dumping of this water before the ship took on cargo.

Adult mitten crabs have been known to travel thousands of kilometres in freshwater at up to 18 km per day. The young crabs, when migrating upriver, seem to be mainly herbivorous. As they grow, they become omnivorous, eating vegetation, crustaceans, insects and dead fish—in fact anything they can get a hold of! Not only is this a problem for the plants and animals that they are eating, but also they compete with native species, such as freshwater crayfish, for food.

(*a*) How does the appearance of the Chinese mitten crab differ from other crabs?

______________________________ **1**

(*b*) State the type of environment the Chinese mitten crabs are found in at each of the following stages in their life.

(i) Early years ______________________________

(ii) Breeding times ______________________________

(iii) Mature adults ______________________________ **1**

(*c*) How is it thought that the Chinese mitten crabs arrived in Britain?

______________________________ **1**

DO NOT WRITE IN THIS MARGIN

Marks KU PS

10. (continued)

(*d*) Describe **one** problem the Chinese mitten crab causes to the habitat and **one** problem it causes to the native community.

Habitat ______________________________ 1

Community ______________________________ 1

(*e*) Describe the changes in its diet as a young adult mitten crab grows.

____

______________________________ 1

(*f*) When moving at their maximum speed, how long would it take an adult mitten crab to travel the whole length of a 45 km river?

Space for calculation

__________ days 1

[Turn over

DO NOT WRITE IN THIS MARGIN

Marks KU PS

11. (*a*) Lactic acid is a waste product from one type of respiration. What type of respiration produces lactic acid?

____________________________ **1**

(*b*) The lactic acid content of the blood of a professional cyclist was measured while cycling at different speeds.

The graph shows the results of these measurements taken at the start of the racing season and at the end.

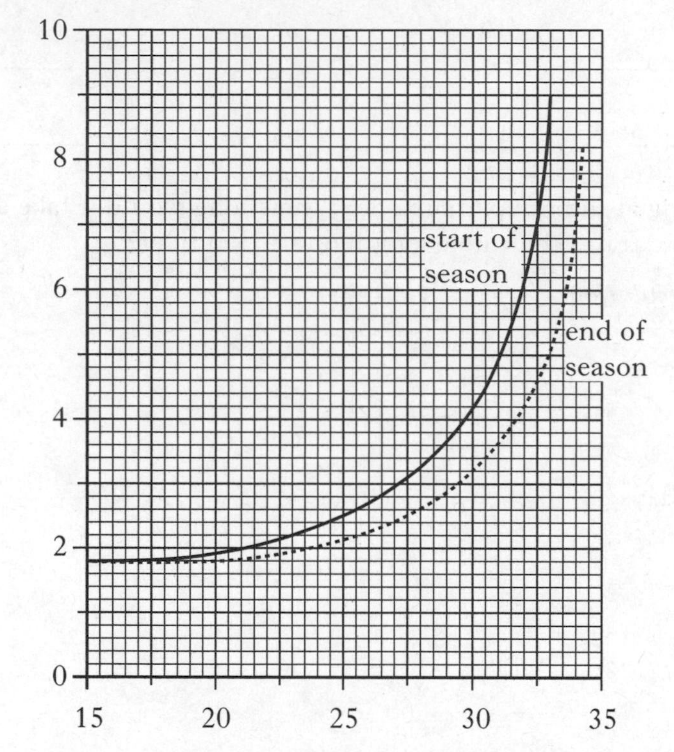

(i) What was the lactic acid concentration when the cyclist was travelling at 15 miles per hour?

_______ nM **1**

(ii) At the start of the season, what was the speed of the cyclist when he was producing 50% of his maximum lactic acid concentration?

_______ miles per hour **1**

DO NOT WRITE IN THIS MARGIN

Marks KU PS

11. (*b*) (continued)

(iii) When lactic acid concentration rises above 2·5 nM, the leg muscles quickly lose power and become painful.

1 What name is given to this condition?

____________________________ **1**

2 What is the maximum speed this cyclist could maintain at the start of the season?

______ miles per hour **1**

(iv) The graph shows that training improves the efficiency of muscles. Other than muscle, name **two** organs whose efficiency is improved by training.

1 __________________________

2 __________________________ **1**

[Turn over

DO NOT WRITE IN THIS MARGIN

Marks	KU	PS

12. Tongue-rolling is an inherited characteristic. The diagram below shows the pattern of its inheritance in one family.

male roller

male non-roller

female roller

female non-roller

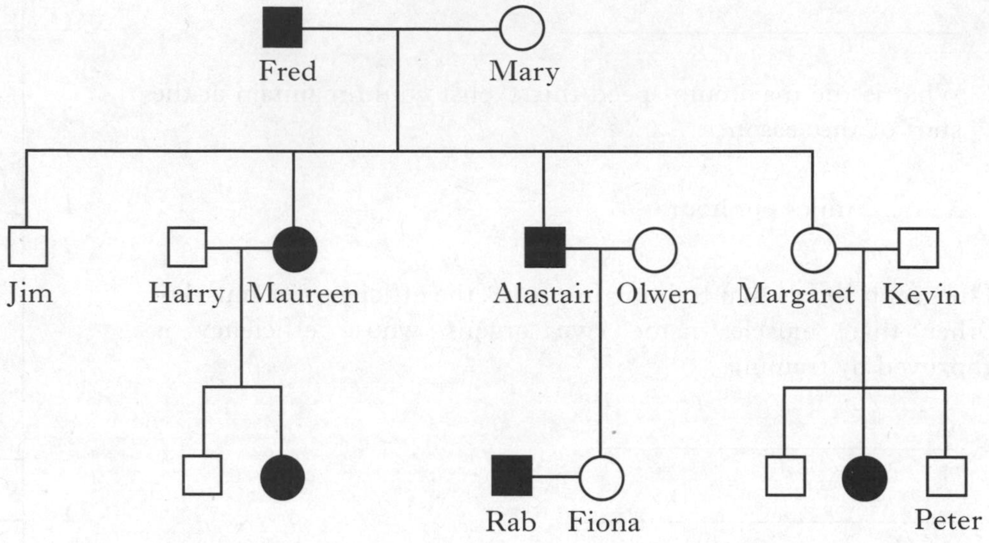

Fred Mary

Jim Harry Maureen Alastair Olwen Margaret Kevin

Rab Fiona Peter

(*a*) (i) Using **R** for the dominant form of the gene and **r** for the recessive form, state the genotypes of:

1 Maureen ______

2 Jim ______

3 Kevin ______ **2**

(ii) If Rab and Fiona have a child, what are the chances of the child being able to roll its tongue?

Space for working

______________________ **1**

(iii) Which of the original parents could be described as true-breeding?

Tick (✓) the correct box.

Fred ☐ Mary ☐

Both ☐ Neither ☐ **1**

(iv) Name a tongue-roller from the F_1 generation.

______________________ **1**

DO NOT WRITE IN THIS MARGIN

12. (continued) *Marks* KU PS

(*b*) Explain why the proportions of the offspring phenotypes from genetic crosses are not always exactly as predicted.

____

____ **1**

(*c*) What term is used for the different forms of the same gene?

____ **1**

[Turn over

DO NOT WRITE IN THIS MARGIN

Marks KU PS

13. The diagram shows an industrial fermenter. It is fitted with a number of taps which allow substances to be added or removed.

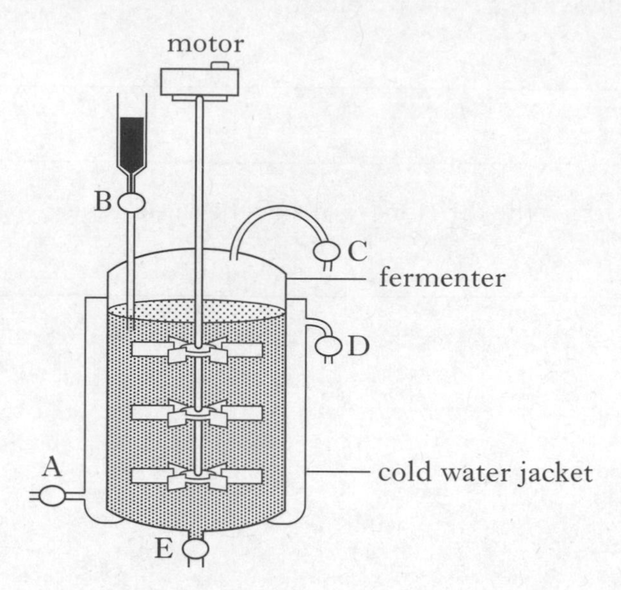

(*a*) Which of the taps, A, B, C, D or E, would open to

(i) add nutrients to the mixture? ______

(ii) remove waste gases? ______

(iii) drain off the products? ______ **2**

(*b*) The fermenter should be kept at 35 °C. Explain why the water jacket around the fermenter should be cold.

____

____ **1**

(*c*) After fermentation is complete, the fermenter is drained and the useful product is separated. New starting ingredients can then be added to the fermenter.

(i) What name is given to this type of process?

______________________ **1**

(ii) When the vessel is empty, it is treated to destroy residual spores of fungi and bacteria. How could this be done?

____

____ **1**

DO NOT WRITE IN THIS MARGIN

Marks	KU	PS

13. (continued)

(*d*) Barley malt extract, water and yeast were placed in the fermenter and left for several days.

The rate of fermentation was measured and the results are shown in the graph below.

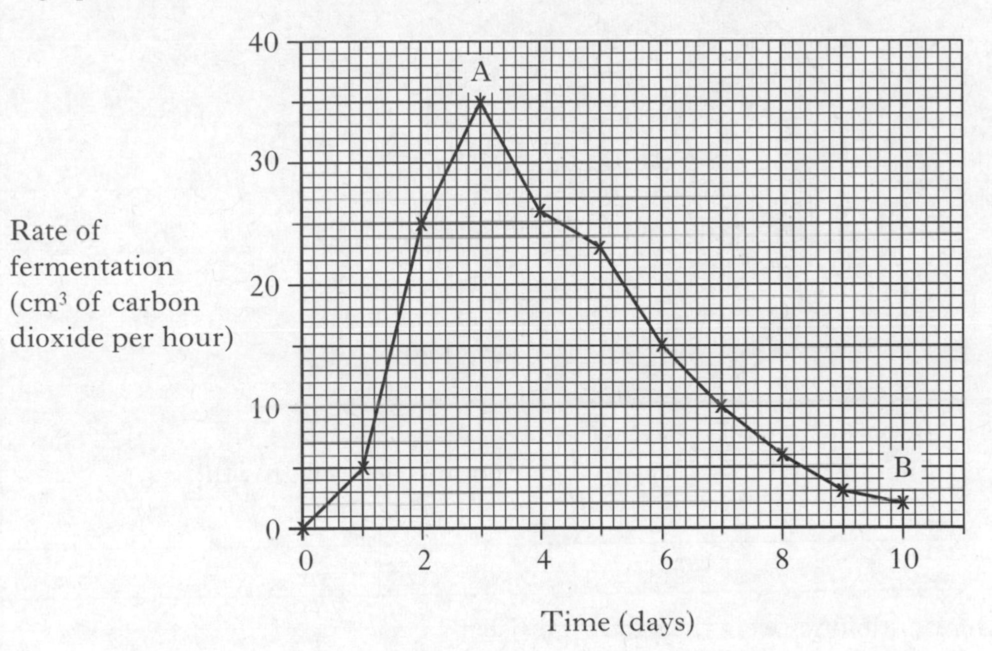

(i) Describe the changes in the rate of fermentation over the ten days.

____

____ **2**

(ii) Suggest a reason for the change in the rate of fermentation between points A and B.

____ **1**

(iii) Why must the barley be malted before it can be used by the yeast?

____

____ **1**

[Turn over for Question 14 on *Page twenty-four*

DO NOT WRITE IN THIS MARGIN

	Marks	KU	PS

14. A nutrient agar plate was covered evenly with a suspension of bacteria. A multidisc was placed on the surface of the agar. Each of the six ends of the multidisc contained a different antibiotic.

The diagram shows the agar plate after incubation.

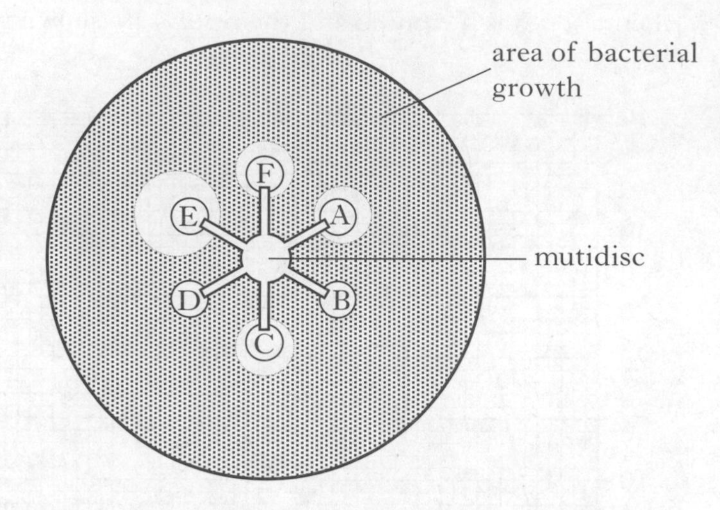

(*a*) Which antibiotic was most effective at preventing bacterial growth?

____________________ **1**

(*b*) To which antibiotics were the bacteria resistant?

____________________ **1**

(*c*) Explain the need for a range of antibiotics in the treatment of diseases caused by bacteria.

____

____ **1**

[*END OF QUESTION PAPER*]

ADDITIONAL GRAPH PAPER FOR QUESTION 4(*a*)

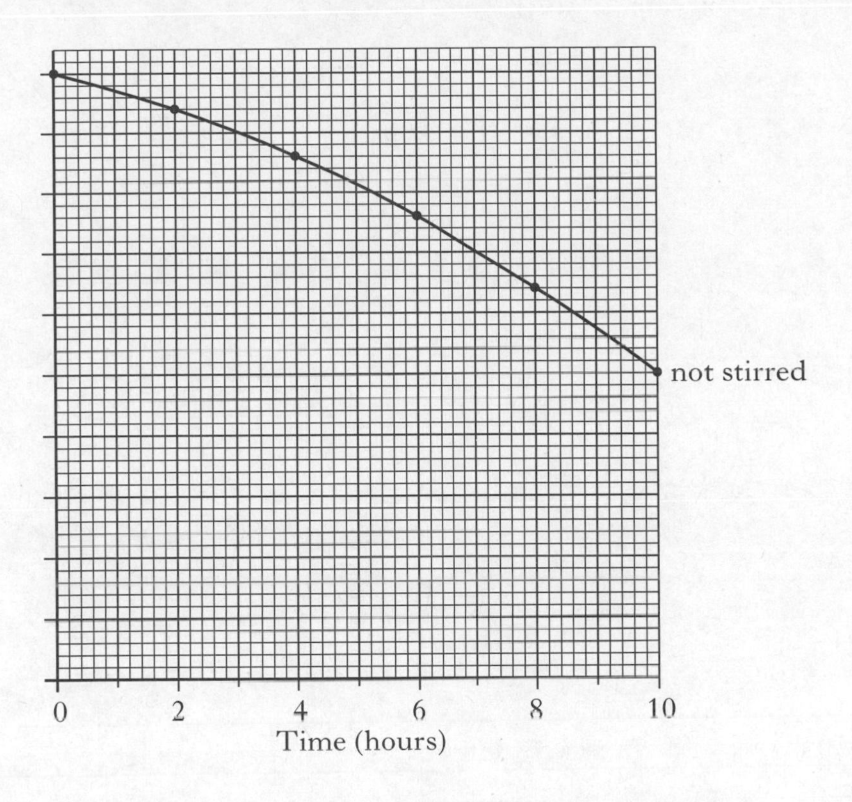

[Turn over

SPACE FOR ANSWERS
AND FOR ROUGH WORKING

STANDARD GRADE | CREDIT

2008

[BLANK PAGE]

FOR OFFICIAL USE

C

KU	PS

Total Marks

0300/402

NATIONAL QUALIFICATIONS 2008

TUESDAY, 27 MAY
10.50 AM – 12.20 PM

BIOLOGY
STANDARD GRADE
Credit Level

Fill in these boxes and read what is printed below.

Full name of centre

Town

Forename(s)

Surname

Date of birth
Day Month Year

Scottish candidate number

Number of seat

1 All questions should be attempted.

2 The questions may be answered in any order but all answers are to be written in the spaces provided in this answer book, and must be written clearly and legibly in ink.

3 Rough work, if any should be necessary, as well as the fair copy, is to be written in this book. Additional spaces for answers and for rough work will be found at the end of the book. Rough work should be scored through when the fair copy has been written.

4 Before leaving the examination room you must give this book to the invigilator. If you do not, you may lose all the marks for this paper.

LI 0300/402 6/30670

DO NOT WRITE IN THIS MARGIN

Marks KU PS

1. (*a*) A comparison was made between the types of invertebrate animals living on the branches and leaves on an oak tree with those living on a beech tree.

Samples were collected as shown below.

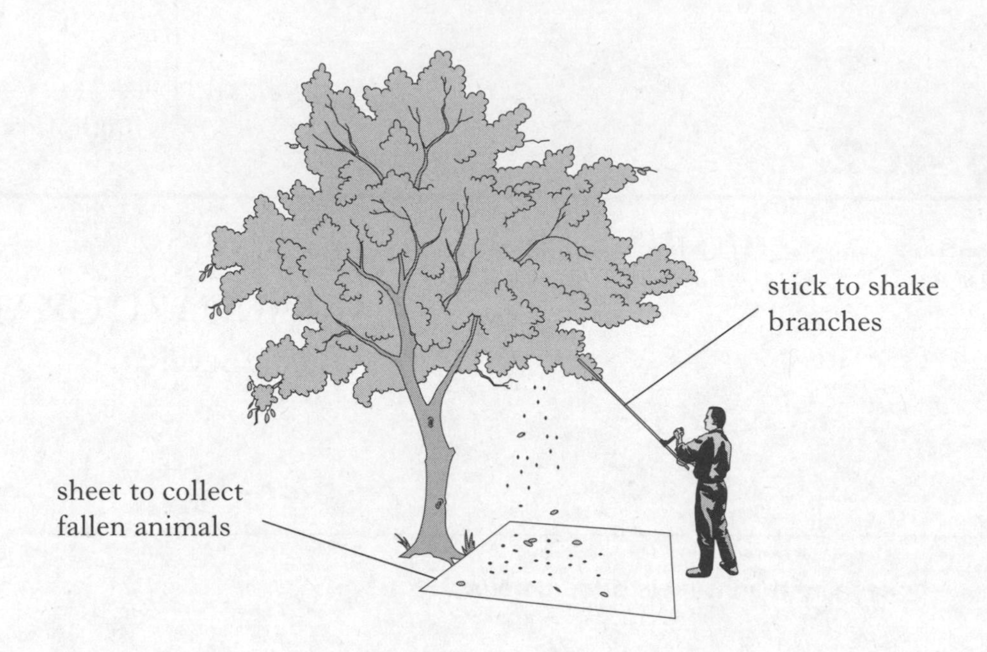

(i) Give **two** variables which should be kept constant to make the comparison valid when using this technique.

1 ______________________________

2 ______________________________ **1**

(ii) The samples collected were not representative of all the invertebrates living on the trees. Suggest a reason for this.

______________________________ **1**

(iii) Measurement of abiotic factors such as light intensity may be recorded at the same time as sampling. Identify a possible source of error for a **named** measurement technique and explain how it might be minimised.

Measurement technique ______________________________

Source of error ______________________________

How to minimise it ______________________________

______________________________ **1**

1. (continued)

(*b*) An investigation was carried out into the effect of light intensity on the distribution of a plant species. At eight different measurement points in a garden, the average light intensity was measured and the percentage ground cover of the plant was recorded.

The results are shown below.

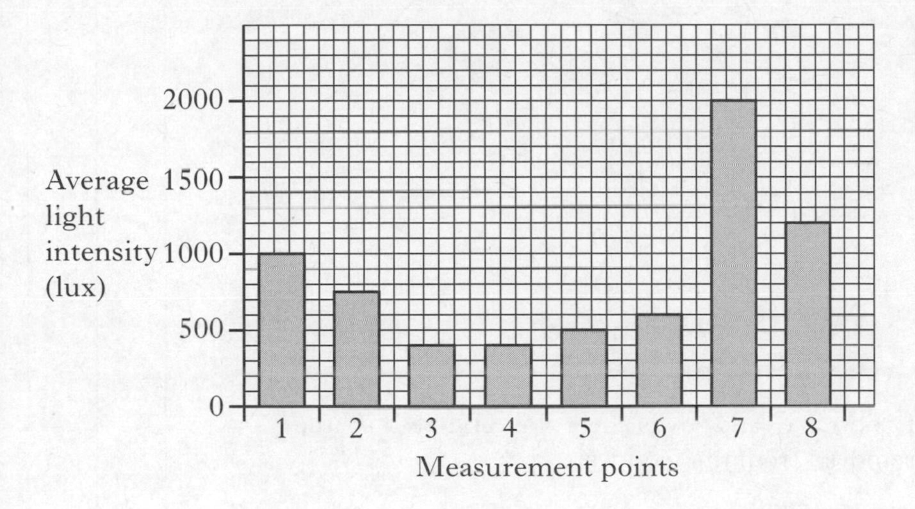

Measurement points	*Ground cover of the plant* (%)
1	85
2	65
3	20
4	20
5	30
6	35
7	100
8	90

DO NOT WRITE IN THIS MARGIN

Marks KU PS

(i) 1 What is the light intensity in the garden where the ground cover of the plant was 100%?

________ lux **1**

2 What was the percentage ground cover of the plant when the light intensity was 750 lux?

________ % **1**

(ii) What is the relationship between light intensity and percentage ground cover of the plant?

____

____ **1**

(*c*) Explain how light intensity affects the distribution of the plants in the garden.

____

____ **1**

DO NOT WRITE IN THIS MARGIN

Marks KU PS

2. (*a*) The diagram shows part of a food web from a forest.

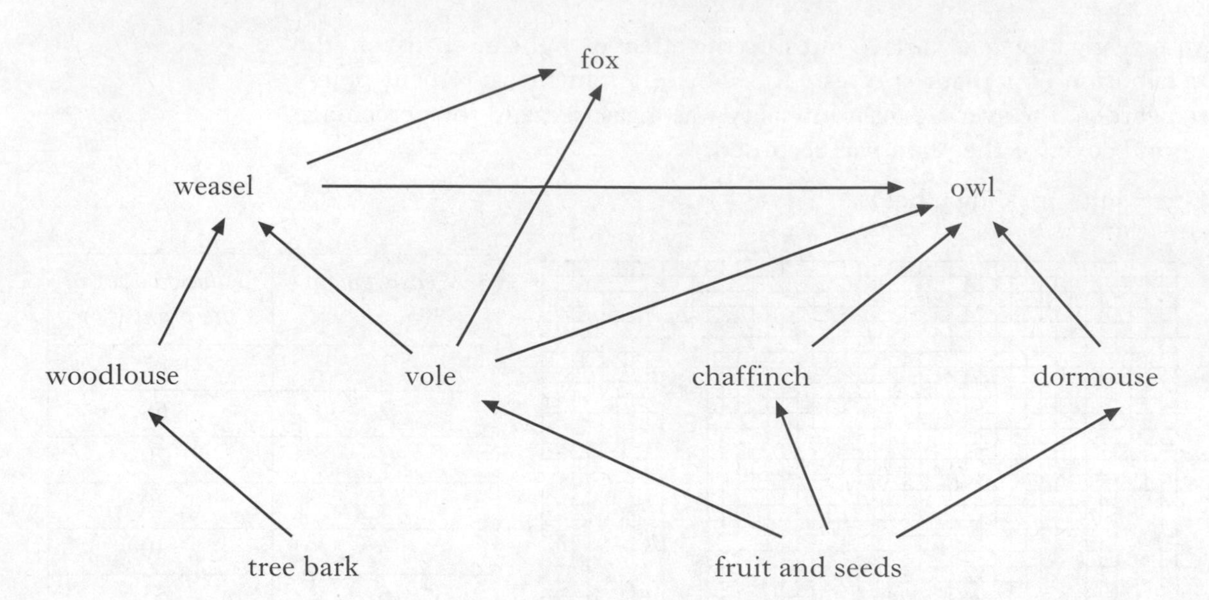

(i) The numbers of dormice and owls may be affected if the chaffinches were removed from the food web.

1 Underline **one** answer in the brackets and give an explanation.

The dormouse population would {increase / decrease / stay the same}.

Explanation ______________________________

______________________________ **1**

2 Underline **one** answer in the brackets and give an explanation.

The owl population would {increase / decrease / stay the same}.

Explanation ______________________________

______________________________ **1**

(ii) Select a food chain from the web which is made up of four stages.

________ → ________ → ________ → ________ **1**

DO NOT WRITE IN THIS MARGIN

Marks	KU	PS

2. (continued)

(*b*) A food chain from the ocean is shown below.

plankton → krill → blue whale

Which population in the food chain has the smallest biomass?

______________________________ **1**

[Turn over

DO NOT WRITE IN THIS MARGIN

Marks KU PS

3. (*a*) The grid contains the names of some components of food.

carbon A	hydrogen B	amino acids C
nitrogen D	simple sugar E	glycerol F
fatty acids G	oxygen H	water I

Use letters from the grid to identify the following:

(i) The sub-units of protein molecules ________ **1**

(ii) The sub-units of fat molecules ________ and ________ **1**

(iii) An element found in protein but not in starch ________ **1**

(*b*) Name the structures in the small intestine which provide an increased surface area for absorption.

____ **1**

(*c*) Urea is produced in the liver from the breakdown of digested food molecules. From which food molecules is urea produced?

____ **1**

DO NOT WRITE IN THIS MARGIN

Marks KU PS

4. (*a*) The diagram shows part of the human breathing system.

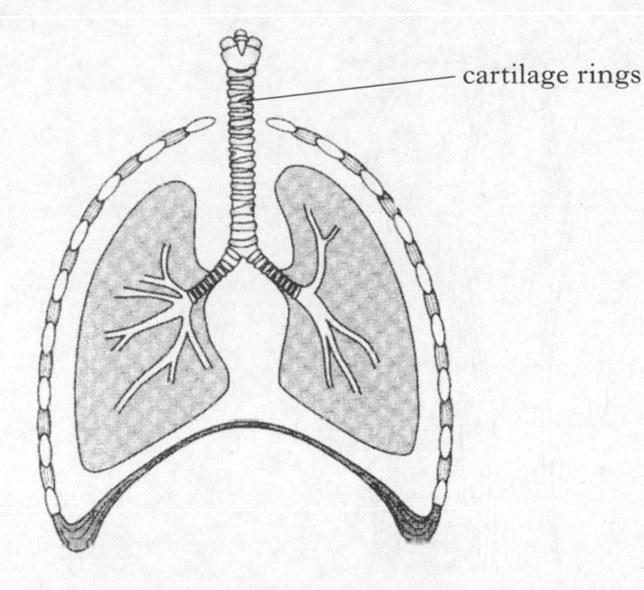

Describe the function of the cartilage rings.

____ 1

(*b*) (i) Name the sticky substance that traps inhaled dust particles.

____________________ 1

(ii) Explain how the trapped particles are removed from the breathing system.

____

____ 1

(*c*) As blood passes through capillary networks in the lungs, oxygen and carbon dioxide are exchanged between the blood and the air sacs.

(i) Describe **one** feature of a capillary network which allows efficient gas exchange.

____

____ 1

(ii) Name the structures in blood that contain haemoglobin.

____________________ 1

(iii) Explain the function of haemoglobin in the transport of oxygen.

____ 1

DO NOT WRITE IN THIS MARGIN

	Marks	KU	PS

5. (*a*) The diagram represents phloem tissue from the stem of a plant.

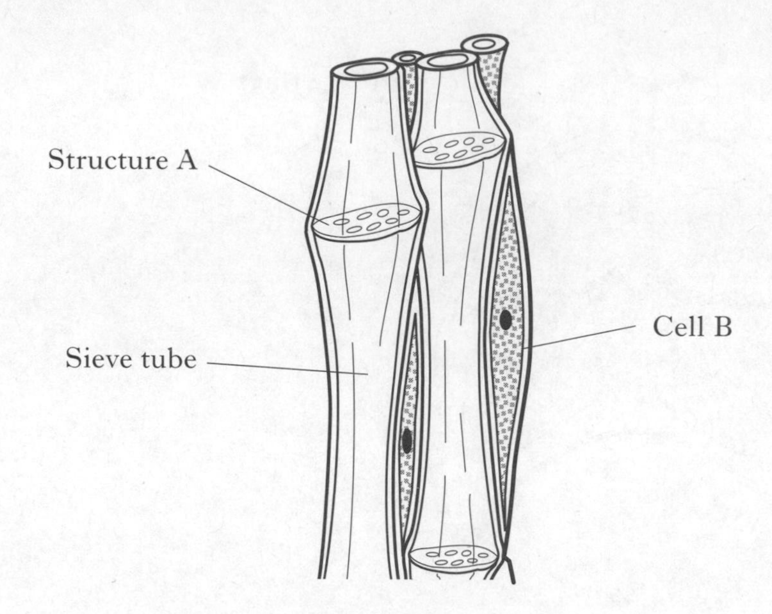

(i) Name Structure A and Cell B.

Structure A ____________________

Cell B ____________________ **2**

(ii) State the function of phloem.

____ **1**

(*b*) (i) Name the leaf tissue where stomata are found.

____________________________ **1**

(ii) Name the cells which control the opening and closing of stomata.

____________________________ **1**

DO NOT WRITE IN THIS MARGIN

Marks KU PS

5. (continued)

(*c*) Leaves were placed in tubes as shown below.

The tubes were left in bright light.

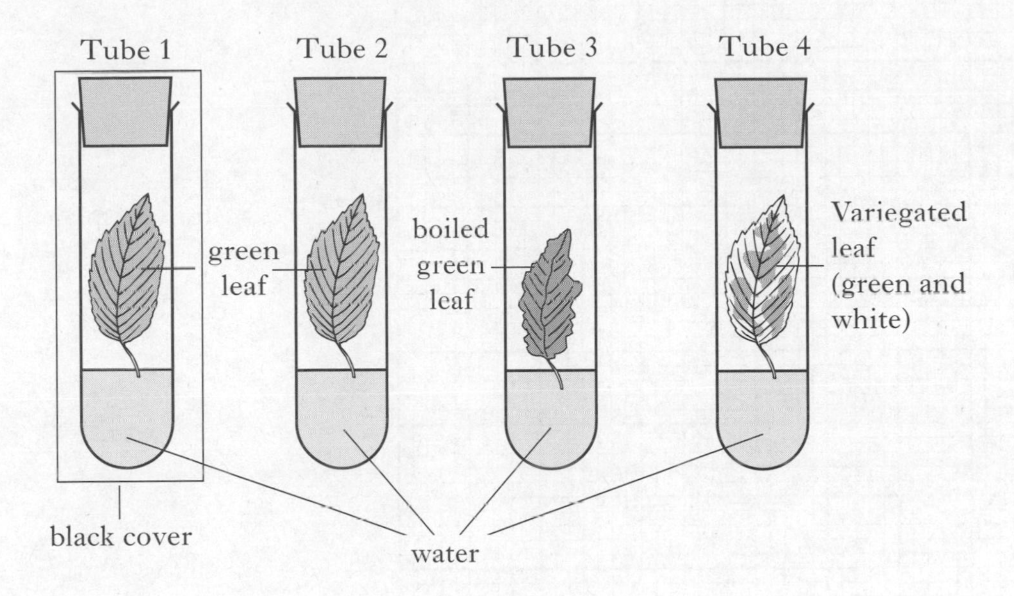

For each of the tubes, tick (✓) the appropriate box in the table to indicate which processes will take place in the leaves.

Process / *Tube*	*Only photosynthesis*	*Only respiration*	*Both*	*Neither*
1				
2				
3				
4				

2

[Turn over

DO NOT WRITE IN THIS MARGIN

Marks KU PS

6. (*a*) The graph shows the number of kidney transplants carried out and the number of patients waiting for a transplant in the UK between 1996 and 2005.

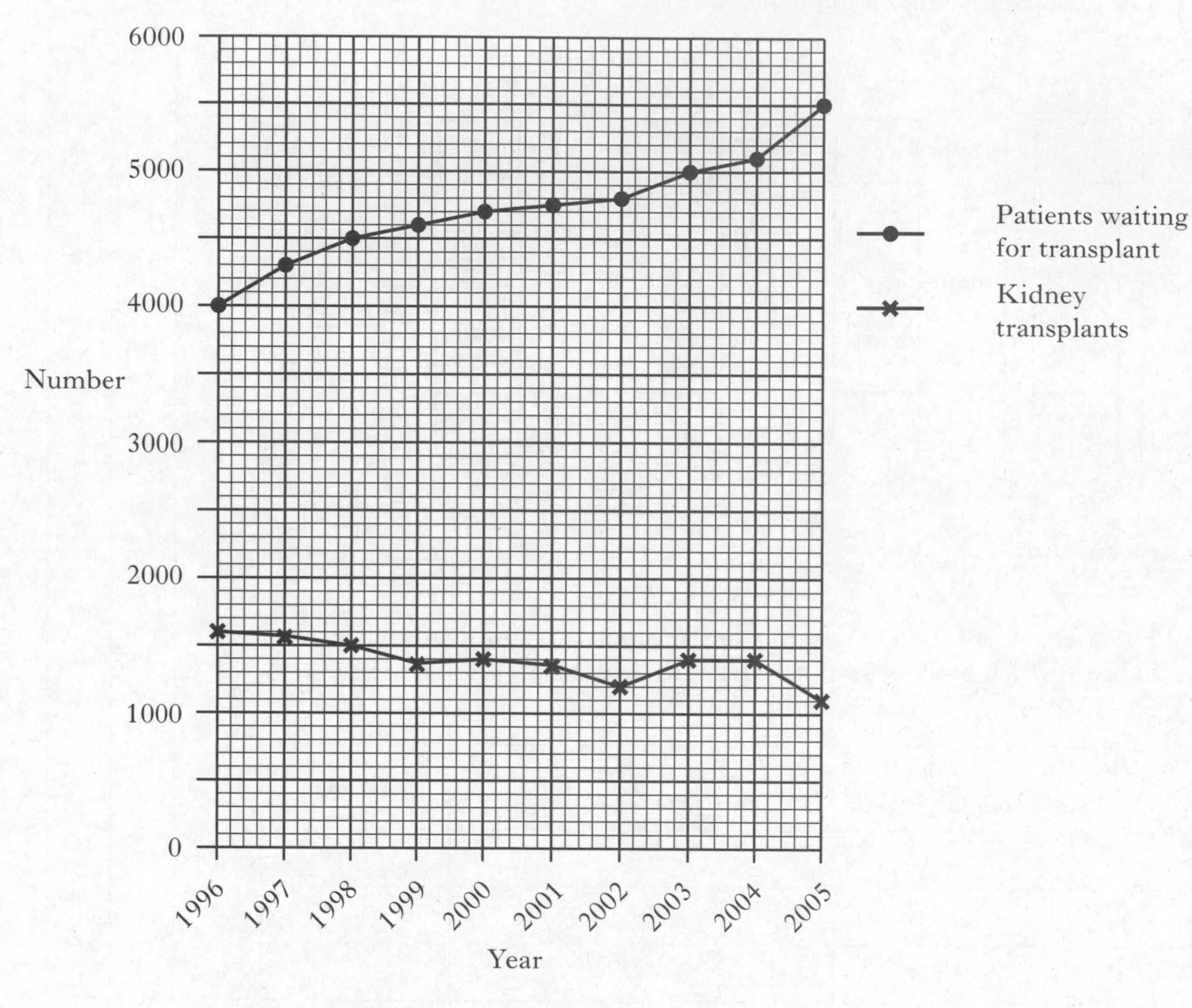

(i) Calculate the average yearly increase in the number of patients waiting for a transplant from 2000 to 2005.

Space for calculation.

Average yearly increase ________ patients per year. **1**

DO NOT WRITE IN THIS MARGIN

Marks KU PS

6. (*a*) (continued)

(ii) Calculate the simple whole number ratios of patients waiting for a transplant to the number of kidney transplants carried out for 1996 and for 2005.

Space for calculation.

1996 ______________ : ______________

2005 ______________ : ______________ **1**

patients waiting for a transplant — transplants carried out

(iii) The following statements refer to the data in the graph.

Tick (✓) the box(es) of the correct statement(s).

The number of patients waiting for a transplant increased every year. ☐

The number of transplants carried out decreased every year. ☐

The difference between the number of patients waiting for a transplant and the number of transplants carried out increased every year. ☐ **1**

(*b*) Give **one** advantage and **one** disadvantage of treating kidney failure by transplant compared to treatment using a dialysis (kidney) machine.

Advantage ____

____ **1**

Disadvantage ____

____ **1**

[Turn over

DO NOT WRITE IN THIS MARGIN

Marks	KU	PS

7. An investigation was carried out into the effect of the mineral boron on the growth of young trout.

Immediately after fertilisation, trout eggs were placed in distilled water containing different concentrates of boron.

After hatching, young trout survive on food from their yolk sac for a maximum of four weeks. The graph below shows the average lengths of the young trout three weeks after hatching.

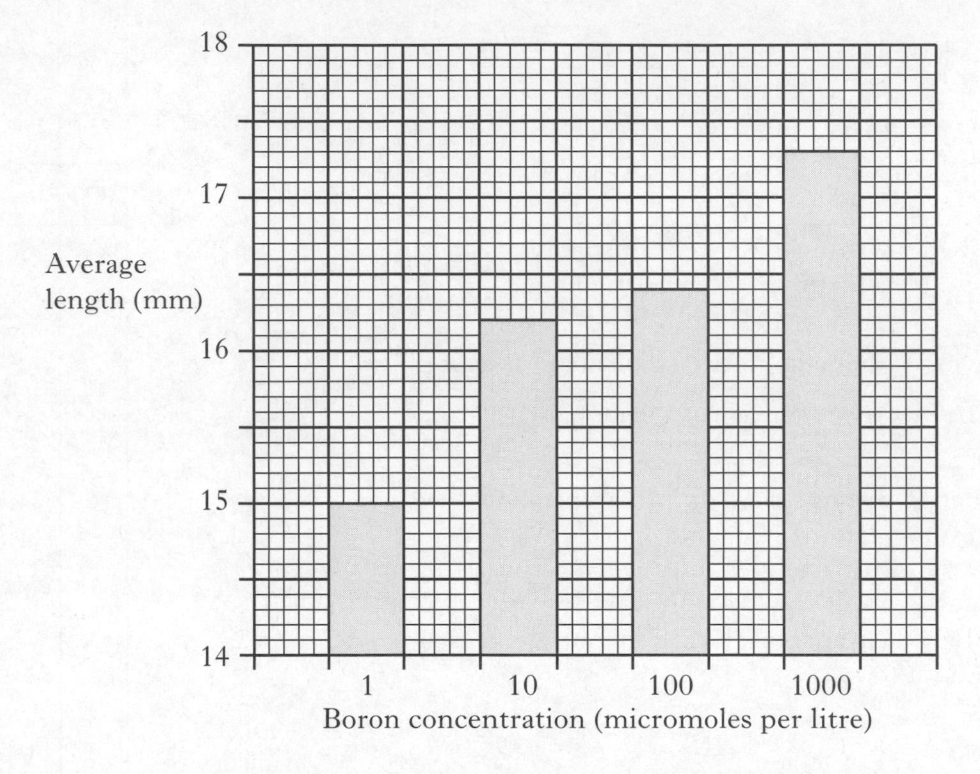

(*a*) Describe the relationship between boron concentration and the length of the young trout.

____

____ **1**

(*b*) Calculate the percentage change in the average fish length when the boron concentration is increased from 1 micromole per litre to 10 micromoles per litre.

Space for calculation.

__________ % **1**

DO NOT WRITE IN THIS MARGIN

Marks	KU	PS

7. (continued)

(*c*) Distilled water is the purest form of water available. Give a reason for using distilled water in this investigation.

____

____ **1**

(*d*) Explain why the results would not be valid if the fish were measured more than four weeks after hatching.

____

____ **1**

[Turn over

DO NOT WRITE IN THIS MARGIN

KU | PS

8. An investigation was carried out into the effect of water concentration on the rate of osmosis.

Details of the apparatus, method used and results are given below.

Apparatus

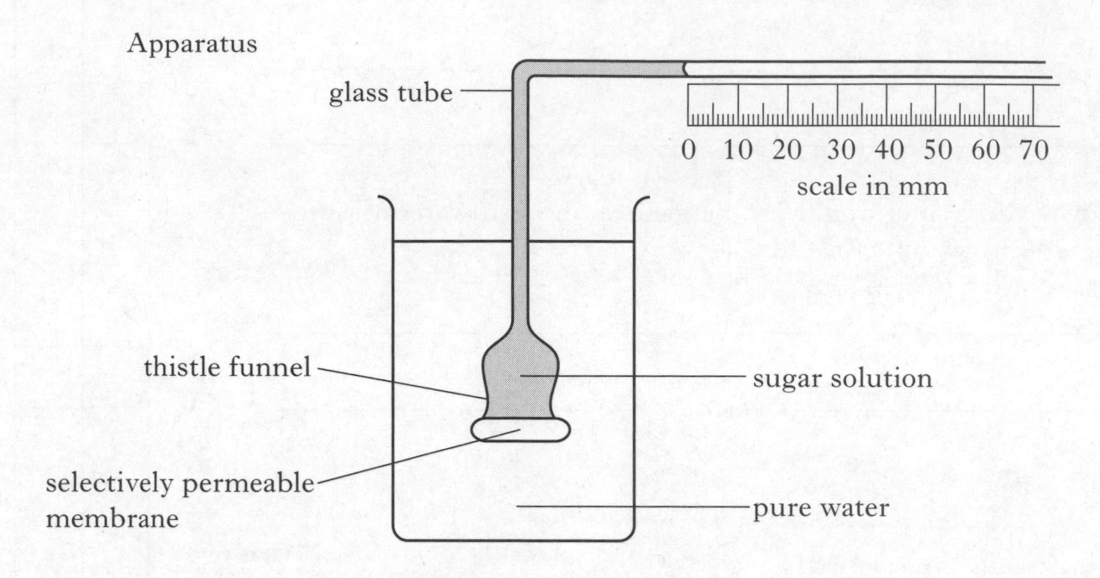

Method

1 A thistle funnel containing 50 cm^3 of 0·5% sugar solution was covered with selectively permeable membrane.
2 The funnel was placed in a beaker of pure water.
3 The scale was positioned with the sugar solution at zero on the scale.
4 The position of the sugar solution was recorded after 30 minutes.
5 The procedure was repeated using 1·0%, 2·0% and 3·0% sugar solutions.

Results

Concentration of sugar solution (%)	*Distance moved by sugar solution in 30 minutes* (mm)
0·5	4·5
1·0	9·0
2·0	18·0
3·0	27·0

DO NOT WRITE IN THIS MARGIN

Marks KU PS

8. (continued)

(*a*) Identify **two** variables not already mentioned that should be kept constant when setting up the investigation.

1 ______________________________

2 ______________________________ **2**

(*b*) Explain the movement of the sugar solution in terms of water concentrations.

____

______________________________ **1**

(*c*) From the results, predict the distance moved by a 3·5% sugar solution in 30 minutes and justify your prediction.

Prediction __________ mm **1**

Justification ______________________________

______________________________ **1**

[Turn over

DO NOT WRITE IN THIS MARGIN

Marks KU PS

9. (*a*) The diagram below contains some of the stages of cell division by mitosis.

Describe **Stages 2** and **5** in the spaces provided.

Stage 1

Chromosomes become visible as pairs of identical chromatids.

↓

Stage 2

1

↓

Stage 3

The spindle fibres contract pulling the chromatids of each chromosome to opposite poles of the cell.

↓

Stage 4

A nuclear membrane forms around each nucleus.

↓

Stage 5

1

(*b*) Mitosis ensures that all daughter cells in a multicellular organism have the same number and type of chromosomes.

Explain why this is necessary.

____

____ 1

DO NOT WRITE IN THIS MARGIN

Marks KU PS

10. (*a*) Barley is a plant grown for use in the brewing industry. The photographs below show two varieties of barley that have been produced by selective breeding.

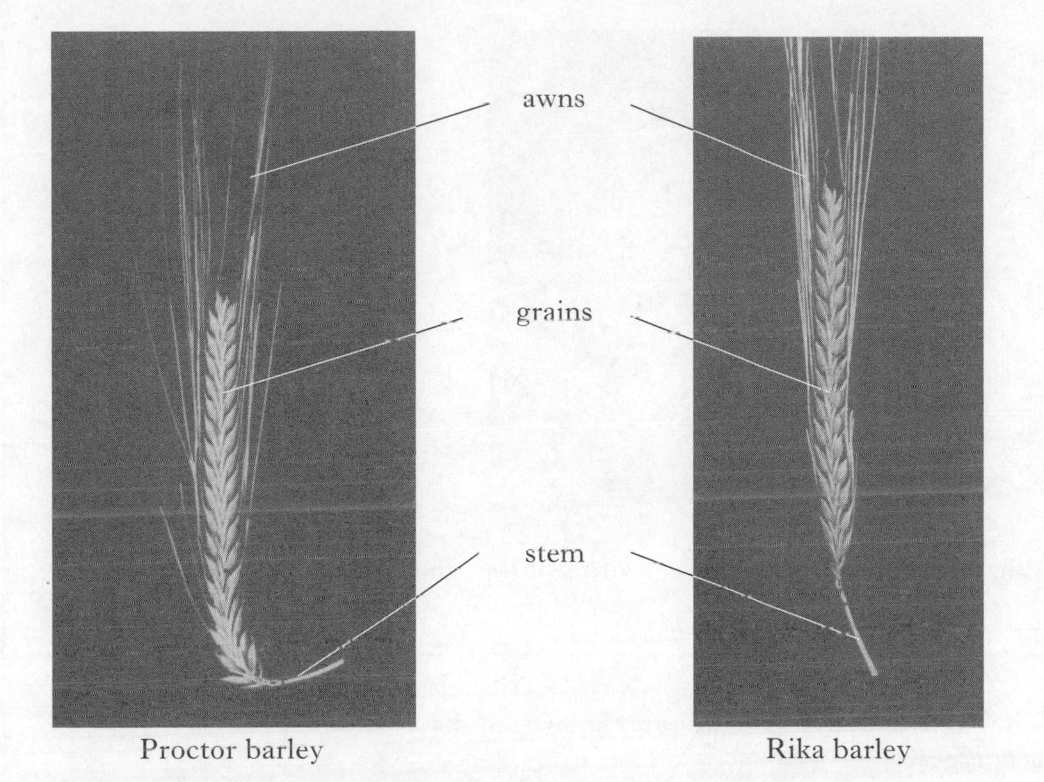

Proctor barley Rika barley

Describe **one** difference between these two varieties of barley.

____

____ 1

(*b*) (i) Explain why barley must be malted before use in the brewing process.

____

____ 1

(ii) Describe how brewers ensure that the yeast carries out fermentation on the sugars extracted from the malted barley.

____ 1

[Turn over

DO NOT WRITE IN THIS MARGIN

Marks KU PS

11. (*a*) The photograph shows a child with dimples. Dimples are small indentations in the cheeks. Their presence is controlled by a single gene which has two forms. The dominant form (**D**) gives dimples. The recessive form (**d**) gives no dimples.

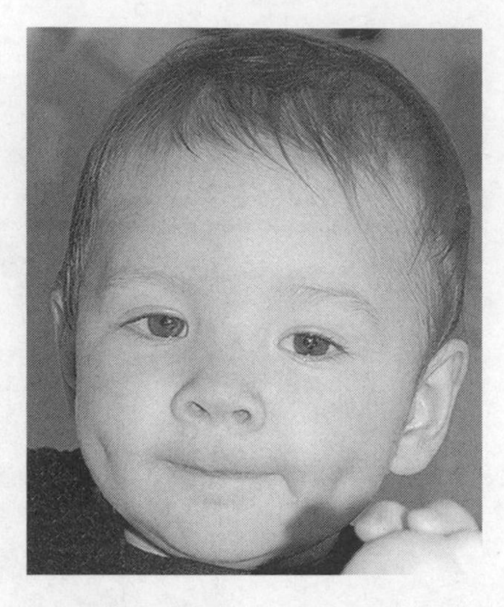

(i) What name is given to different forms of the same gene?

____ 1

(ii) The parents of the child are known to have the following genotypes.

DD × **dd**

Underline **one** option in each bracket to make the following sentence correct.

The parents have {the same / different} phenotypes and {the same / different} genotypes. 1

(iii) What is the genotype of this child?

____ 1

DO NOT WRITE IN THIS MARGIN

Marks KU PS

11. (continued)

(*b*) The diagram shows a cross between tall and dwarf pea plants.

P **Tall** × **Dwarf**

↓

$\mathbf{F_1}$ all **Tall**

↓

$\mathbf{F_2}$ some **Tall**, some **Dwarf**

(i) What would be the predicted ratio of **Tall** to **Dwarf** plants in the $\mathbf{F_2}$ generation?

______ : ______

Tall **Dwarf** **1**

(ii) The observed ratio of **Tall : Dwarf** plants was different from the expected ratio.

Give an explanation for this difference.

______________________________ **1**

(iii) Identify the true-breeding plants from the above cross.

Tick (✓) the box(es) of the correct plant(s).

Tall P ☐

Dwarf P ☐

Tall $\mathbf{F_1}$ ☐ **1**

[Turn over

DO NOT WRITE IN THIS MARGIN

Marks	KU	PS

12. An investigation was carried out into the effect of temperature on the rate of respiration by yeast.

Details of the apparatus, method used and results are given below.

Apparatus

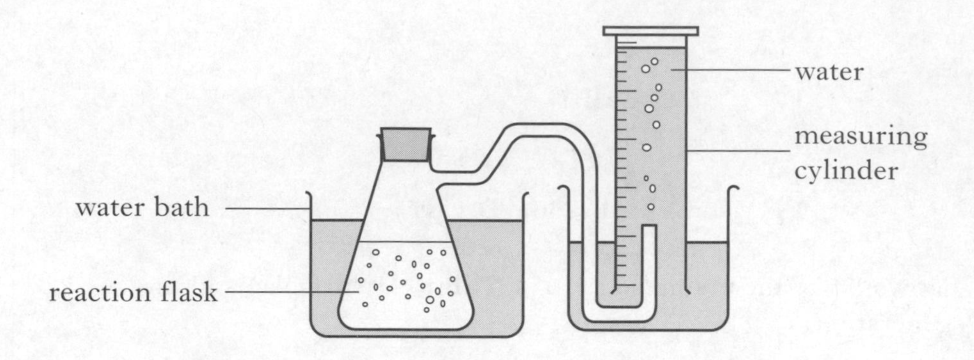

Method

1 Water baths were set up over a range of temperatures.
2 $100\,cm^3$ of glucose solution and $50\,cm^3$ of yeast suspension were allowed to reach the same temperature as the water bath.
3 The glucose solution and the yeast suspension were mixed in the reaction flask.
4 After 1 hour, the volume of gas in the measuring cylinder was measured.

Results

Temperature (°C)	10	20	30	40	50
Volume of gas produced in 1 hour (cm^3)	9	18	36	48	5

(*a*) Ethanol was formed in the reaction flask.

What cell process produced this?

____ **1**

(*b*) Describe the relationship between the temperature and the volume of gas produced in one hour.

____

____ **2**

DO NOT WRITE IN THIS MARGIN

Marks	KU	PS

12. (continued)

(*c*) Predict the volume of gas which would be collected in one hour if the investigation was repeated at 60 °C. Give an explanation for your answer.

Prediction ________ cm^3 **1**

Explanation ____

____ **1**

(*d*) Describe the control flasks that would be set up to show that the gas was produced due to activity of the yeast and to no other factor.

____

____ **2**

(*e*) Use the results to complete a line graph to show the volumes of gas produced in one hour over the range of temperatures.

(An additional grid, if needed, will be found on page 27.)

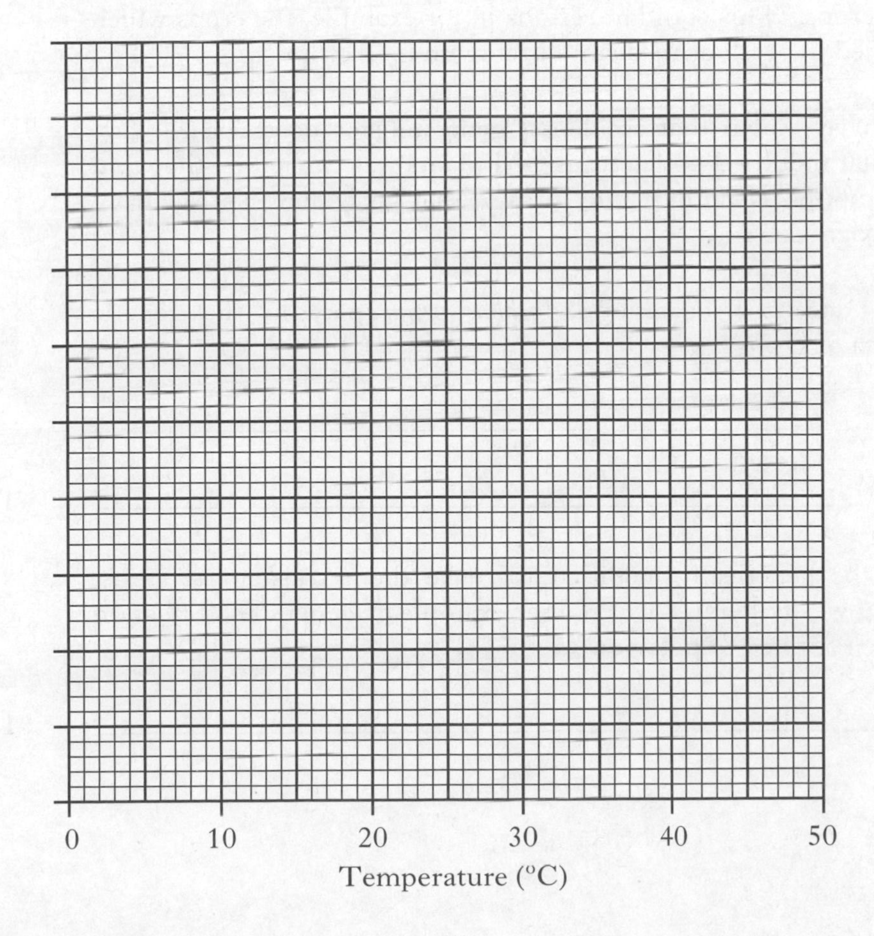

2

[Turn over

DO NOT WRITE IN THIS MARGIN

Marks | KU | PS

13. Read the following passage and answer the questions based on it.

Adapted from ***GM Organisms*** by John Pickrell, www.newscientist.com

Genetic modification (GM) of crops began with the discovery that the soil bacterium *Agrobacterium* could be used to transfer useful genes from unrelated species into plants. The Bt gene is one of the most commonly inserted. It produces a pesticide toxin that is harmless to humans but is capable of killing insect pests. Many new crop types have been produced. Most of these are modified to be pest, disease or weedkiller resistant, and include wheat, maize, oilseed rape, potatoes, peanuts, tomatoes, peas, sweet peppers, lettuce and onions.

Supporters argue that drought resistant or salt resistant varieties can flourish in poor conditions. Insect-repelling crops protect the environment by minimising pesticide use. Golden rice with extra vitamin A or protein-enhanced potatoes can improve nutrition.

Critics fear that GM foods could have unforeseen effects. Toxic proteins might be produced or antibiotic-resistance genes may be transferred to human gut bacteria. Modified crops could become weedkiller resistant "superweeds". Modified crops could also accidentally breed with wild plants or other crops. This could be serious if, for example, the crops which had been modified to produce medicines bred with food crops.

Investigations have shown that accidental gene transfer does occur. One study showed that modified pollen from GM plants was carried by the wind for tens of kilometres. Another study proved that genes have spread from the USA to Mexico.

(*a*) What role does the bacterium *Agrobacterium* play in the genetic modification of crops?

____

____ **1**

(*b*) Crops can be genetically modified to make them resistant to pests, diseases and weedkillers. Give another example of genetic modification that has been applied to potatoes.

____ **1**

DO NOT WRITE IN THIS MARGIN

Marks KU PS

13. (continued)

(*c*) Explain why a plant, which is modified to be weedkiller resistant could be:

(i) useful to farmers.

____

____ 1

(ii) a problem for farmers.

____

____ 1

(*d*) Give **one** example of a potential threat to health by the use of GM crops.

____

____ 1

[Turn over

DO NOT WRITE IN THIS MARGIN

Marks KU PS

14. (*a*) In a commercial process, a bacterial species is provided with glucose and produces a hormone. The bacteria release the hormone into surrounding liquid. The graph shows changes in the glucose concentration and the hormone concentration during a 60 hour period.

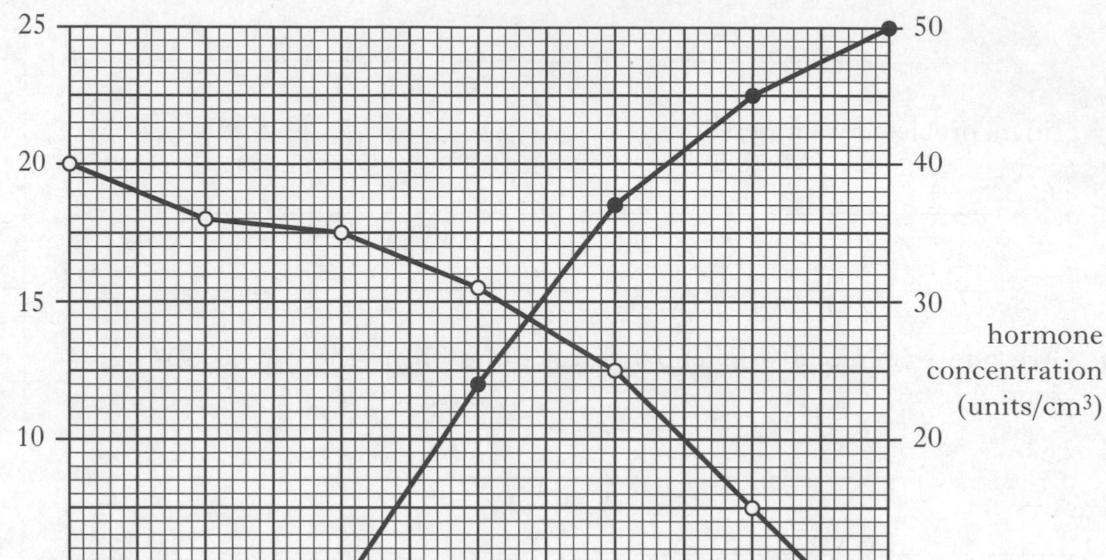

(i) How many hours did it take for 50% of the glucose to be used up by the bacteria?

__________ hours **1**

(ii) During which 10 hour period was secretion of hormone the greatest?

Tick (✓) the correct box.

☐ 20 – 30 hours

☐ 30 – 40 hours

☐ 40 – 50 hours

☐ 50 – 60 hours **1**

DO NOT WRITE IN THIS MARGIN

Marks KU PS

14. (a) (continued)

(iii) Calculate the decrease in glucose concentration over the 60 hour period.

Space for calculation.

________ g/100 cm^3 **1**

(iv) If glucose continues to be used at the same rate as between 50 and 60 hours, predict how many more hours it would be before all the glucose would be used up.

Space for calculation.

________ hours **1**

(v) During the first 10 hours of the process, energy was being used for functions other than the synthesis of the hormone.

Give **two** pieces of evidence from the graph to support this statement.

1 ________________________________

2 ________________________________ **1**

(b) Glucose is a carbohydrate component of food. Which food component contains most energy per gram?

________________________________ **1**

[Turn over for Question 15 on *Page twenty-six*

DO NOT WRITE IN THIS MARGIN

Marks KU PS

15. (*a*) In a sewage works, micro-organisms cause the decay of the sewage. What is the benefit to the micro-organisms in carrying out this process?

______________________________ 1

(*b*) What type of respiration must be carried out by the micro-organisms to ensure complete breakdown of the sewage?

______________________________ 1

(*c*) Sewage contains a wide range of materials. What ensures that all these materials are broken down?

______________________________ 1

(*d*) The table shows the methods of disposal of the sludge obtained from sewage treatment.

Method of disposal of sludge	*Percentage*
Spread on farmland	50
Landfill	10
Dumped at sea	15
Incinerated	20
Other disposal	5

Use the information from the table to complete the pie chart below.

(An additional chart, if needed, will be found on page 27.)

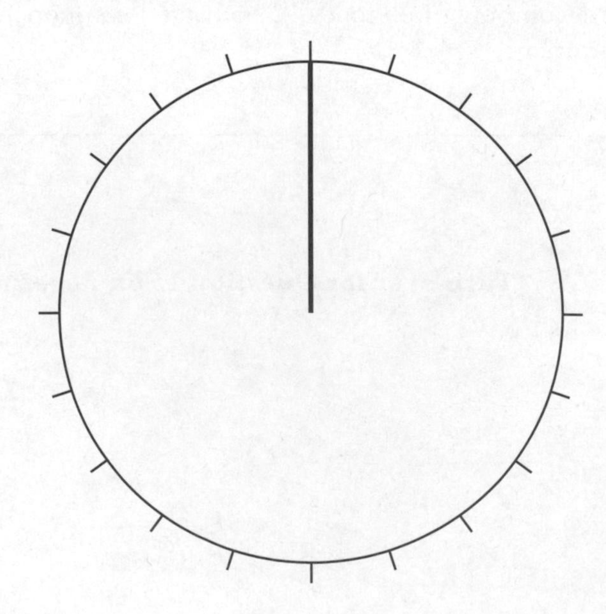

2

[*END OF QUESTION PAPER*]

ADDITIONAL GRAPH PAPER FOR QUESTION 12(*e*)

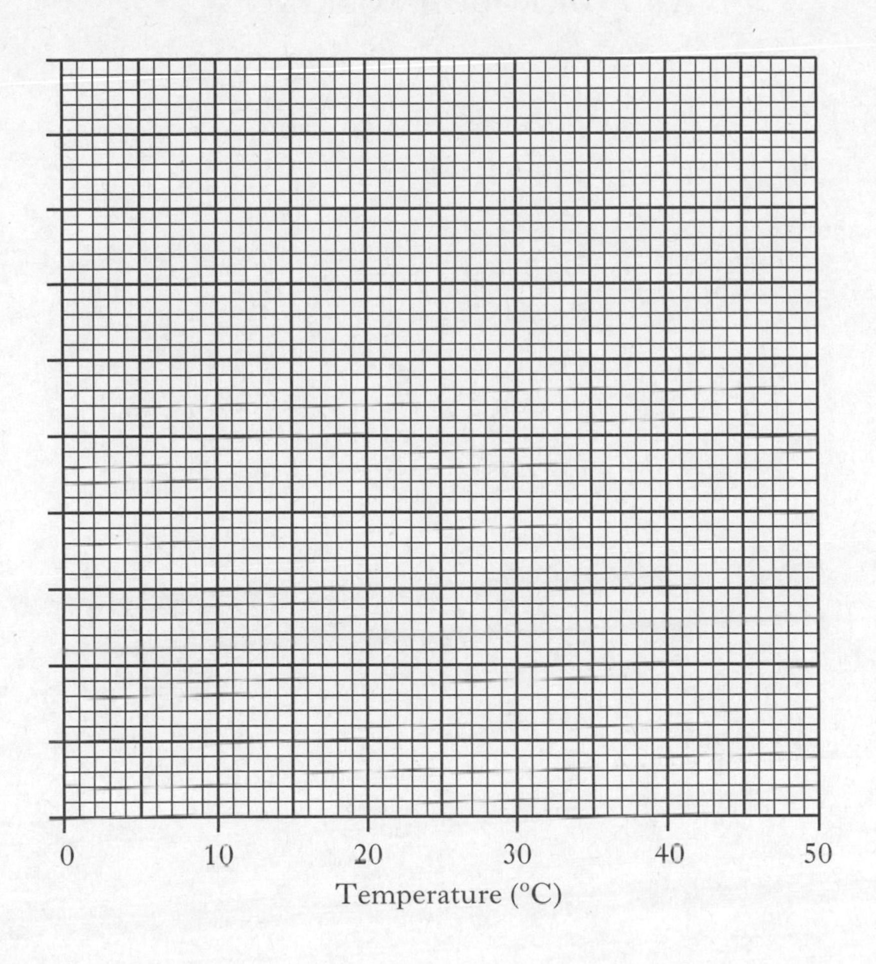

ADDITIONAL PIE CHART FOR QUESTION 15(*d*)

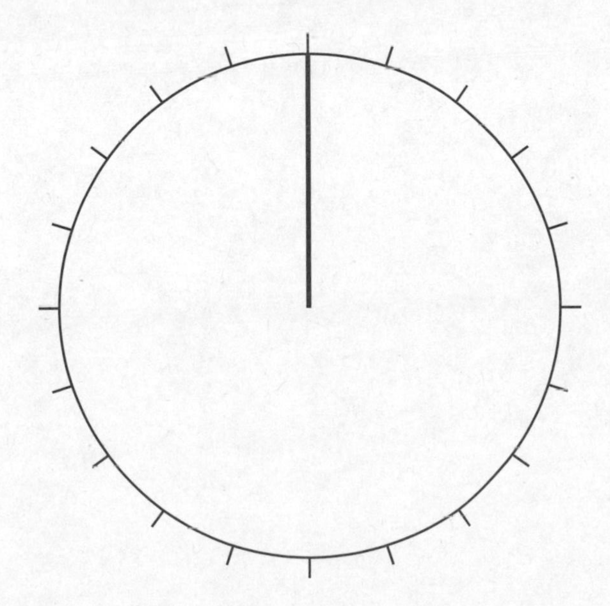

SPACE FOR ANSWERS
AND FOR ROUGH WORKING

STANDARD GRADE | CREDIT

2009

[BLANK PAGE]

FOR OFFICIAL USE

C

KU	PS

Total Marks

0300/402

NATIONAL QUALIFICATIONS 2009

THURSDAY, 28 MAY
10.50 AM – 12.20 PM

BIOLOGY
STANDARD GRADE
Credit Level

Fill in these boxes and read what is printed below.

Full name o f centre

Town

Forename(s)

Surname

Date of birth
Day Month Year

Scottish candidate number

Number of seat

1 All questions should be attempted.

2 The questions may be answered in any order but all answers are to be written in the spaces provided in this answer book, and must be written clearly and legibly in ink.

3 Rough work, if any should be necessary, as well as the fair copy, is to be written in this book. Additional spaces for answers and for rough work will be found at the end of the book. Rough work should be scored through when the fair copy has been written.

4 Before leaving the examination room you must give this book to the invigilator. If you do not, you may lose all the marks for this paper.

DO NOT WRITE IN THIS MARGIN

Marks KU PS

1. (*a*) Rabbits were first brought to Australia by European settlers.
The graph below shows the change in rabbit population in Australia since their introduction.

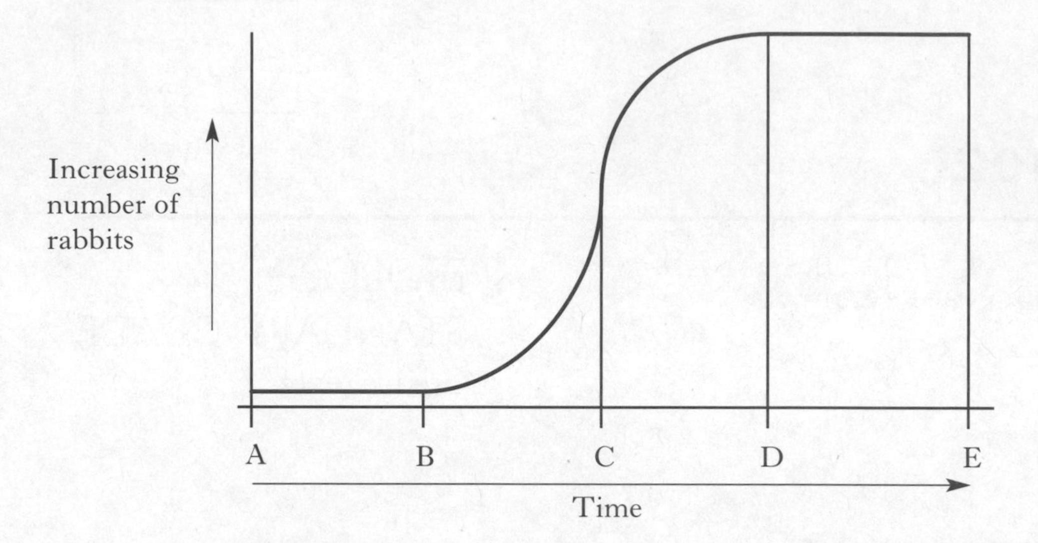

(i) Describe the changes in the rabbit population between times A and E.

____

______________________________ **2**

(ii) Suggest one reason for the population change between times B and C.

____

______________________________ **1**

(*b*) To control over-grazing by rabbits, a disease was introduced in 2005 which was fatal to rabbits but not to other species.
If this disease had wiped out the rabbit population, what effect could it have had on the population of:

(i) Eastern wallabies which are herbivores?

(ii) Dingoes which are carnivorous wild dogs?

Explain your answers.

(i) Effect on Eastern wallabies ______________________________

Explanation ______________________________

(ii) Effect on Dingoes ______________________________

Explanation ______________________________ **2**

DO NOT WRITE IN THIS MARGIN

Marks KU PS

2. (*a*) Coal-burning and nuclear power stations are used to produce electricity in Britain.

Draw lines to connect each type of power station with features considered to be adverse effects of their operation.

Type of power station

coal burning

nuclear

Features

- Waste can cause high levels of acid rain
- Waste must be sealed before it is stored
- High volume of greenhouse gas production
- Waste is dangerous for hundreds of years

2

(*b*) Environmental protection analysis was carried out on water samples from three burns.

The Mains Burn had the highest pH at 8·0. It also had the highest oxygen saturation at 94% compared to Bell's Burn which had the lowest at 65%.

The Hatchery Burn had the lowest value for suspended solids at 4·0 mg/l, with an oxygen saturation of 91·5%.

Bell's Burn had a suspended solids reading of 5·6 mg/l and the lowest pH at 7·7 compared to a value of 7·9 for the Hatchery Burn. The highest reading for suspended solids was recorded in the Mains Burn with a value of 6·0 mg/l.

(i) Complete the following table with the data in the passage using suitable column headings.

Analysis site			
Hatchery Burn			
Bell's Burn			
Mains Burn			

3

(ii) Calcium in the water of the burns raises the pH.

Water snails need calcium for shell growth. Which burn would you expect to have the highest number of water snails?

____

1

DO NOT WRITE IN THIS MARGIN

Marks KU PS

3. (*a*) The diagram below represents a wind-pollinated flower.

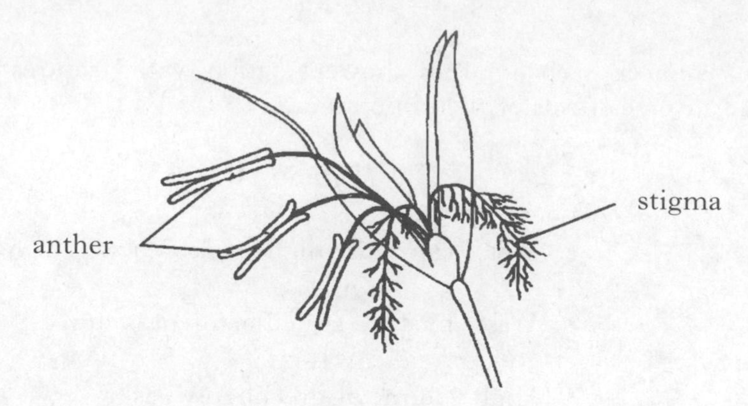

Explain how each of the labelled structures contributes to wind pollination.

Anther ______________________________

____

Stigma ______________________________

______________________________ **2**

(*b*) The chart below shows the peak times for airborne pollen from six wind-pollinated plants.

	Month											
Type of plant	*Jan*	*Feb*	*Mar*	*Apr*	*May*	*Jun*	*Jul*	*Aug*	*Sep*	*Oct*	*Nov*	*Dec*
Hazel												
Yew												
Willow												
Oil seed rape												
Grass												
Nettle												

(i) How many months are shown to be free of pollen?

______________________________ **1**

(ii) The above plants account for most pollen allergy in Britain.

Most allergy sufferers are affected for 3–4 months each year.

Give a conclusion which can be drawn about pollen allergy from these facts.

______________________________ **1**

DO NOT WRITE IN THIS MARGIN

Marks KU PS

3. (*b*) (continued)

(iii) In summer, air carries an average of 100 pollen grains per litre.

If a person inhales 12·6 litres of air per minute, calculate the total number of pollen grains inhaled each hour.

Space for calculation

__________ grains per hour **1**

(*c*) What essential stage in plant reproduction must take place after pollination and before fertilisation?

______________________________________ **1**

(*d*) Give one example of a plant which relies on wind for seed dispersal and describe how its seeds are adapted to dispersal in this way.

Plant ______________________

Description ______________________

______________________________________ **2**

(*e*) The list below describes groups of organisms.

1 a patch of strawberry plants produced from the runners of one plant
2 a field of barley grown from seeds
3 a litter of pedigree West Highland Terrier puppies
4 a group of potato tubers harvested from the same plant
5 all the pea plants grown from peas from the same pod

Use the numbers from the list to identify each of the groups which form a clone.

Numbers ______________________ **1**

[Turn over

DO NOT WRITE IN THIS MARGIN

Marks KU PS

4. (*a*) The table below shows some features of five British butterflies.

Butterfly species	*Wing shading*	*Wing tip*	*Wing spots*
Large White	pale	black	yes
Orange Tip	pale	orange	no
Peacock	dark	blue	yes
Red Admiral	dark	white	yes
Wood White	pale	black	no

Complete the key using the information given in the table.

1 Pale wing shading .... go to 2

Dark wing shading .... [ ]

2 [ ] .... [ ]

Orange wing tip .... **Orange Tip**

3. Spots on wings .... **Large White**

No spots on wings .... [ ]

4. Blue wing tip .... **Peacock**

[ ] .... [ ]

3

DO NOT WRITE IN THIS MARGIN

Marks KU PS

4. (continued)

(*b*) The earliest sighting of these butterflies in Britain was recorded in 1956 and again in 2006. The information is shown in the table below.

	Earliest sighting	
Butterfly species	1956	2006
Large White	mid June	early June
Orange tip	late May	mid May
Peacock	mid March	early March
Red Admiral	early June	late May
Wood White	mid May	early May

(i) What evidence suggests that the average temperatures in 2006 were higher than in 1956?

____

______________________________ 1

(ii) What name is given to organisms, such as these butterflies, which can be used to provide information about environmental factors?

______________________________ 1

[Turn over

DO NOT WRITE IN THIS MARGIN

Marks KU PS

5. (*a*) The table below shows information on the number of eggs fertilised and the survival of offspring for four different animals.

Animal	*Average number of eggs fertilised at one time*	*Average number of surviving offspring*	*Percentage survival rate*
Dog	5	4	
Human	1	1	100
Bird	4	3	75
Trout	1000	20	2

(i) Calculate the percentage survival rate for the dog and complete the table with the result.

Space for calculation

1

(ii) Explain the difference in the survival rates between humans and trout.

____

____

1

(*b*) Embryos of mammals exchange substances with their mother through the placenta.

Name a substance which passes through the placenta from an embryo to its mother.

____

1

DO NOT WRITE IN THIS MARGIN

Marks KU PS

6. The diagram below shows *Paramecium,* a single-celled organism which lives in water.

(*a*) The water concentration outside the cell is higher than the water concentration of the cytoplasm. This causes water to enter the cell constantly.

(i) What is the name for this movement of water?

____________________ **1**

(ii) From the information given, state whether *Paramecium* is likely to live in fresh water or salt water.

____________________ **1**

(*b*) *Paramecium* must get rid of excess water. Pure water is collected in the vacuoles by removing it from the cytoplasm. The vacuoles are emptied to the surrounding water as soon as they are full.

(i) What would happen to the *Paramecium* cell if the vacuoles stopped working properly?

____________________ **1**

(ii) The vacuoles are not filled by the diffusion of water.

What evidence is there to support this statement?

____

____________________ **1**

[Turn over

DO NOT WRITE IN THIS MARGIN

Marks | KU | PS

7. (*a*) Underline one word in each bracket to make the paragraph about water balance correct.

When a large volume of water is taken into the body, the water

concentration of the blood {increases / decreases}. The volume of ADH

released into the blood by the pituitary gland {increases / decreases}.

This causes water reabsorption by the kidneys to {increase / decrease}

and the volume of urine produced increases. **2**

(*b*) The diagram below represents a nephron from a kidney.

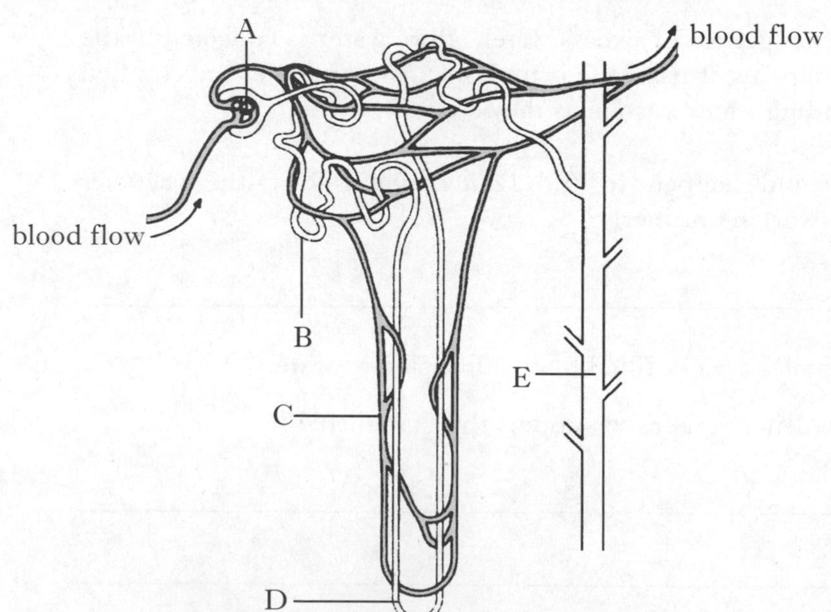

(i) Which letter on the diagram shows where filtration occurs?

__________ **1**

(ii) Which letter on the diagram shows a collecting duct?

__________ **1**

DO NOT WRITE IN THIS MARGIN

7. (continued)

Marks | KU | PS

(*c*) The table below shows the concentration of some substances found in samples taken from the blood, the kidney filtrate and the urine of a volunteer.

Substance	*Concentration in blood* ($g/100cm^3$)	*Concentration in filtrate* ($g/100cm^3$)	*Concentration in urine* ($g/100cm^3$)
urea	0·25	0·25	2·00
glucose	0·10	0·10	0·00
protein	7·50	0·00	0·00
salts	0·62	0·62	1·50

(i) Which substance was present in the blood but was not filtered out of it?

____________________ **1**

(ii) Which substance was filtered from the blood and then completely reabsorbed back into it?

____________________ **1**

(*d*) A person produces an average of 1·8 litres of urine per day and this is 1% of the kidney filtrate.

What is the average volume of filtrate reabsorbed daily?

Space for calculation

____________________ litres **1**

[Turn over

DO NOT WRITE IN THIS MARGIN

Marks KU PS

8. (*a*) Stages of mitosis are shown in their correct order in the diagrams below.

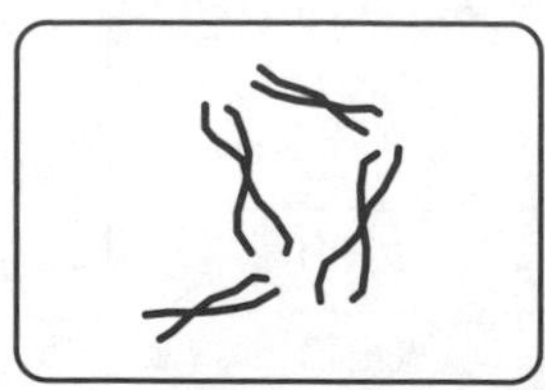

Stage A

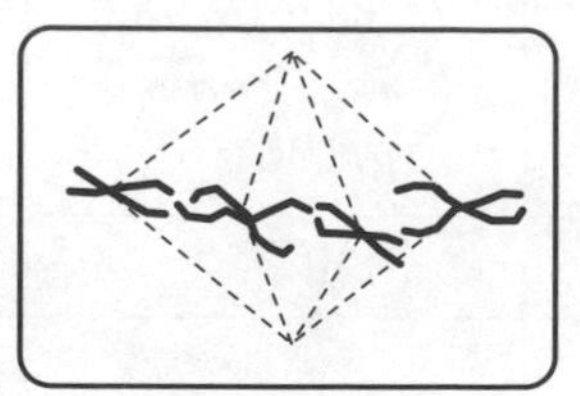

Stage B

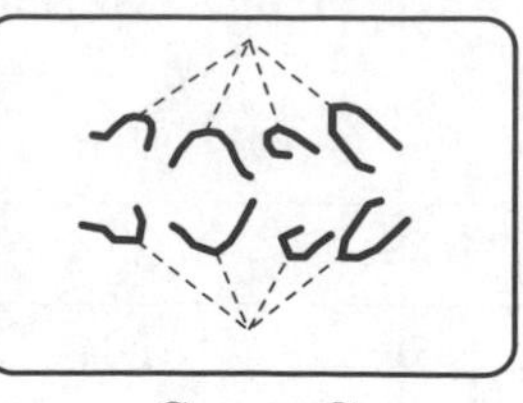

Stage C

(i) Label the spindle on one of the diagrams. **1**

(ii) Stage C would be followed by stage D. Describe what would happen in stage D.

____

____ **1**

(*b*) Typical timings of the stages of mitosis are shown in the table below.

Stage	A	B	C	D
Time (minutes)	88	33	25	54

What percentage of the total time for mitosis is taken by stage C?

Space for calculation

______________ % **1**

(*c*) Scientists can grow liver tissue in the laboratory. This is done by making a few liver cells divide by mitosis to form a large mass of cells.

Why is it important that the daughter cells contain the same number of chromosomes as the original mother cells?

____

____ **1**

DO NOT WRITE IN THIS MARGIN

Marks KU PS

9. (*a*) The diagram below represents a hinge joint.

Complete each of the boxes with the missing name or function of the part labelled.

Name
Function

Name
Synovial fluid
Function

2

(*b*) Tendons attach muscle to bone.

Explain why it is important that tendons are inelastic.

____

____

____

2

[Turn over

DO NOT WRITE IN THIS MARGIN

Marks KU PS

10. (*a*) The following statements refer to breathing.

1 ribs move up and out
2 ribs move down and in
3 diaphragm relaxes
4 diaphragm contracts
5 chest volume decreases
6 chest volume increases
7 lung pressure decreases
8 lung pressure increases

Complete the box by inserting the statement numbers which refer to breathing in.

Statements referring to ***breathing in***

2

(*b*) The table below shows how exercise at different work rates affects heart rate, breathing rate and the lactic acid concentration in the blood.

Work rate (watts)	*Heart rate* (beats/min)	*Breathing rate* (breaths/min)	*Lactic acid concentration* (mg/l)
0	76	12	1·0
40	92	13	1·5
80	112	15	1·8
120	132	16	3·5
160	156	18	4·5
200	172	30	9·0

(i) Calculate the percentage increase in lactic acid concentration as the work rate increases from 0 to 200 watts.

Space for calculation

__________ %

1

DO NOT WRITE IN THIS MARGIN

Marks | KU | PS

10. (*b*) (continued)

(ii) Explain why the lactic acid concentration increases as the work rate increases.

____

____ **1**

(iii) The graph uses information from the table to show how the breathing rate varies with work rate.

On the same grid, add a scale and label to the vertical axis on the **left side** and plot a line graph to show how the heart rate varies with work rate.

(An additional graph, if needed, will be found on *Page twenty-six*.)

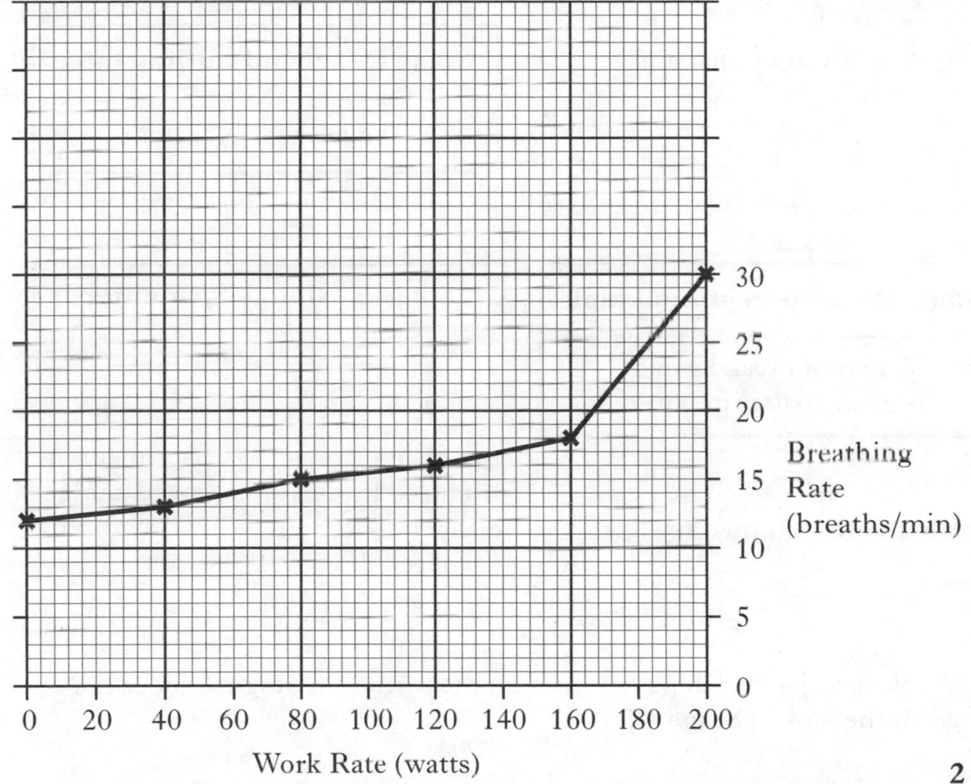

2

(iv) Describe the relationship between work rate and both breathing and heart rates.

____

____ **1**

DO NOT WRITE IN THIS MARGIN

Marks KU PS

11. The flow chart shows what happens in a typical sewage treatment works.

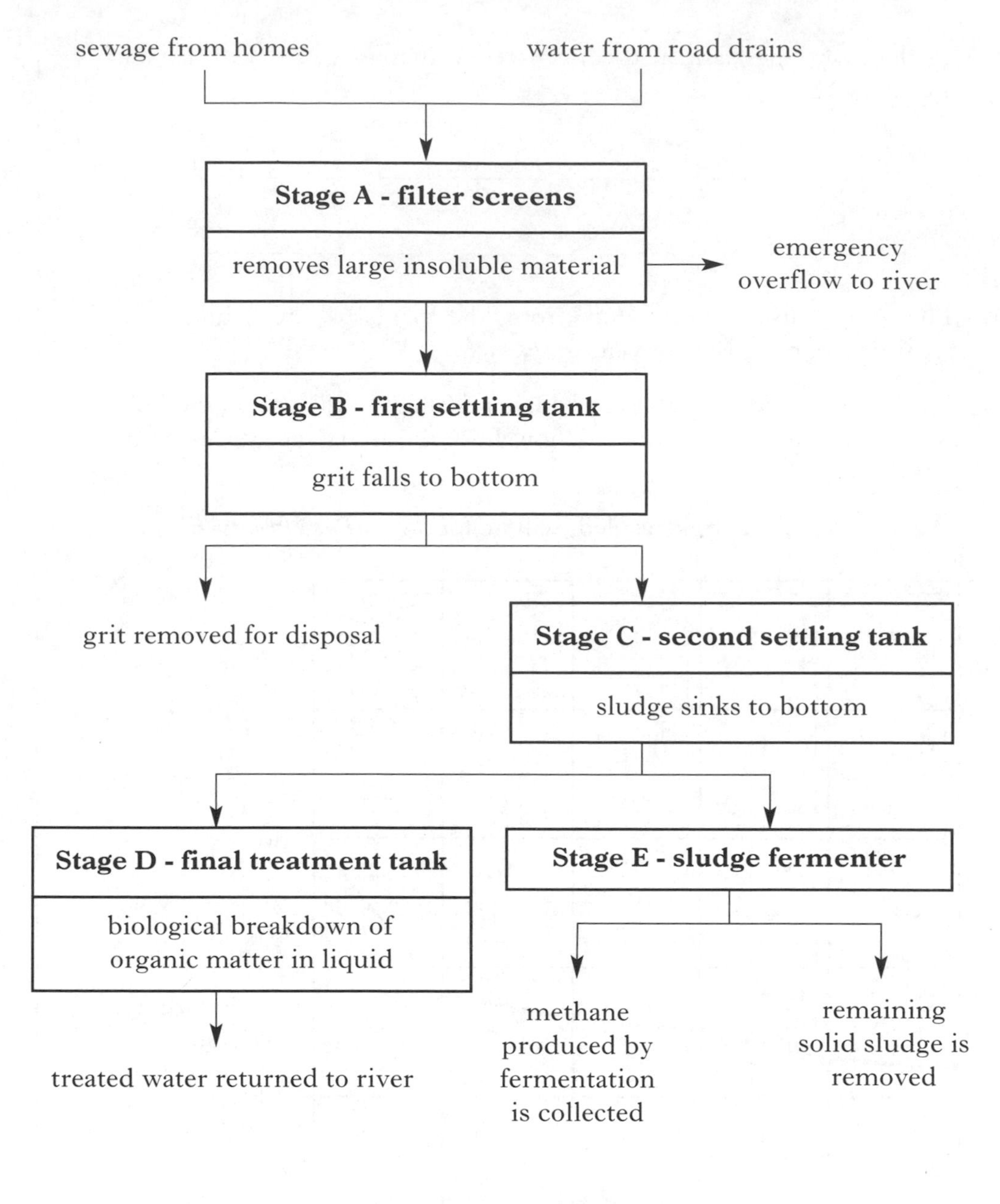

(*a*) What material, which passes through the screens in **Stage A**, does not reach the tank in **Stage C**?

____________________________ **1**

(*b*) Name the gas needed for the final treatment in **Stage D** and explain why the gas is needed for this process.

Gas __________________________

Explanation ____

____ **2**

DO NOT WRITE IN THIS MARGIN

Marks KU PS

11. (continued)

(*c*) When liquid from **Stage D** was sampled, it was found to contain over 80 different species of micro-organisms. Explain why this was seen as a good result.

____

____ 1

(*d*) Under what environmental conditions could untreated sewage enter the river, even if the sewage treatment plant was working correctly?

____ 1

[Turn over

DO NOT WRITE IN THIS MARGIN

Marks | KU | PS

12. Read the following passage and answer the questions using information from it.

Salve Imperator Adapted from "The Life of Birds" by David Attenborough.

Reproduction for Emperor penguins involves extreme hardship. They start their breeding cycle in March at the beginning of the Antarctic winter. At this time the fringe of ice that surrounds the Antarctic continent is at its narrowest. The penguins walk across it for several miles to the permanent ice which is their breeding ground. Up to 25 000 penguins gather and mating takes place in April.

As the temperature falls, the sea ice expands by 2 miles per day. In May the female produces one large egg which she places on the top of her feet. The male takes the egg, juggles it onto the top of his feet and covers it with a fold of his densely feathered abdomen to keep it warm. Producing the egg has taken a significant proportion of the female's body reserves. She needs to replenish them urgently and heads back to sea to feed.

As the winter winds begin to blow, the temperature falls. The male Emperors huddle closer together for warmth and shelter. They use their tiny stump of a tail as the third leg of a tripod and rest on their heels. Their upwardly turned toes keep their precious eggs off the ice. There is nothing to eat and for a month there is total darkness.

After 60 days the eggs hatch. The males, close to starvation, manage to produce a little milky secretion from their gullets for their chicks. At this critical moment the females reappear. They have had a long journey as the ice has extended considerably. The females regurgitate their chicks' first real meal. The males now start the long trek back to the sea to feed for the first time in four months.

Three weeks later, the males are back to take over the care of the chicks, allowing the females to return to the sea. As winter slackens its grip, the ice begins to break up. The journey to the sea gets shorter and the parents can increase the frequency of feeding. In November the parents stop feeding the young and long processions of adults and young waddle down to the sea.

(*a*) Why is it necessary for the females to leave their eggs and return to the sea?

____ **1**

DO NOT WRITE IN THIS MARGIN

Marks | KU | PS

12. (continued)

(*b*) By how much has the distance to the sea increased in the time between laying and hatching?

Space for calculation

____________________ miles **1**

(*c*) How does a male keep his egg off the ice?

____

____ **1**

(*d*) The following list describes events in the life of Emperor penguins.

List
1 walk to breeding grounds
2 mating
3 egg laying
4 eggs hatch and females return
5 parents and chicks waddle to the sea

Complete the time line below by placing the number of each event in the correct month.

(An additional time line will be found, if needed, on *Page twenty-six*).

Time line

Jan	Feb	Mar	Apr	May	Jun	Jul	Aug	Sep	Oct	Nov	Dec

2

(*e*) How many months of the year are **not** spent breeding and rearing young?

__________ **1**

[Turn over

DO NOT WRITE IN THIS MARGIN

Marks	KU	PS

13. The following apparatus was used to investigate the effectiveness of washing powders.

Identical pieces of stained cloth were washed using different washing powders.

The cloths were dried and the degree of stain removal was measured by recording light reflected from the cloth with a light meter. The meter was set to read 100% when the cloth was perfectly clean. Any stain left on the cloth reduced the intensity of light recorded.

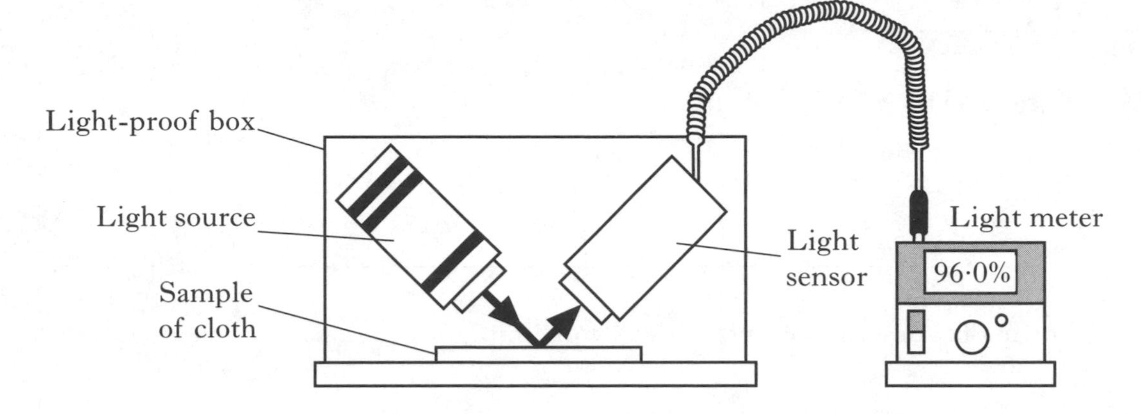

(*a*) (i) Various precautions were taken to ensure that the experimental procedure was valid.

Identify the point(s) which contributed to this.

Tick (✓) the correct box(es).

The procedure used gave appropriate information about the effectiveness of washing powders. ☐

All significant variables were controlled and were identical except the one being investigated. ☐

Several results were collected and used to calculate an average. ☐ **1**

(ii) Explain why it was necessary to carry out the investigation in a light-proof box.

____

____ **1**

DO NOT WRITE IN THIS MARGIN

Marks | KU | PS

13. (continued)

(*b*) The results obtained using two different washing powders at various temperatures are shown below.

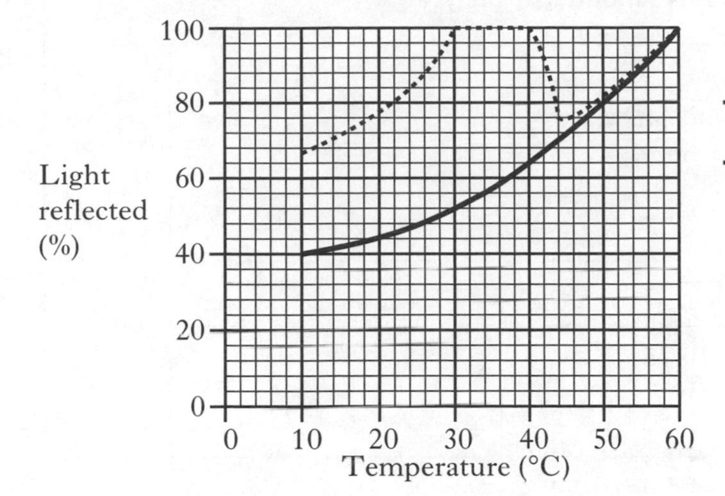

(i) At which temperature was there the greatest difference between the effectiveness of the two washing powders?

________ °C **1**

(ii) Each one degree Celcius reduction in the washing temperature saves 2p in the cost of electricity used to heat the water for each wash.

Calculate the annual saving in the electricity costs to achieve 100% stain removal with biological washing powder compared to a non-biological one, for a household which does one wash per week.

Space for calculation

annual saving = £________ **1**

(iii) What type of biological substance gives biological washing powders their properties?

________________ **1**

(iv) Explain why the effectiveness of the biological washing powder decreases between 40°C and 45°C.

________________________________ **1**

[Turn over

DO NOT WRITE IN THIS MARGIN

Marks KU PS

14. Micro-organisms living in water use dissolved oxygen for respiration.

The mass of oxygen they use is called the Biochemical Oxygen Demand (BOD).

The table below shows the BOD of a river and the concentration of solid material carried by the river during five months of the year.

Month	*Concentration of solid material* (mg/l)	*BOD* (mg/l)
January	6·75	1·0
March	7·25	1·2
May	10·75	1·9
September	5·50	0·5
November	9·00	1·5

(*a*) Use the information in the table to complete the bar chart below for January and November.
(An additional chart, if needed, will be found on *Page twenty-seven.*)

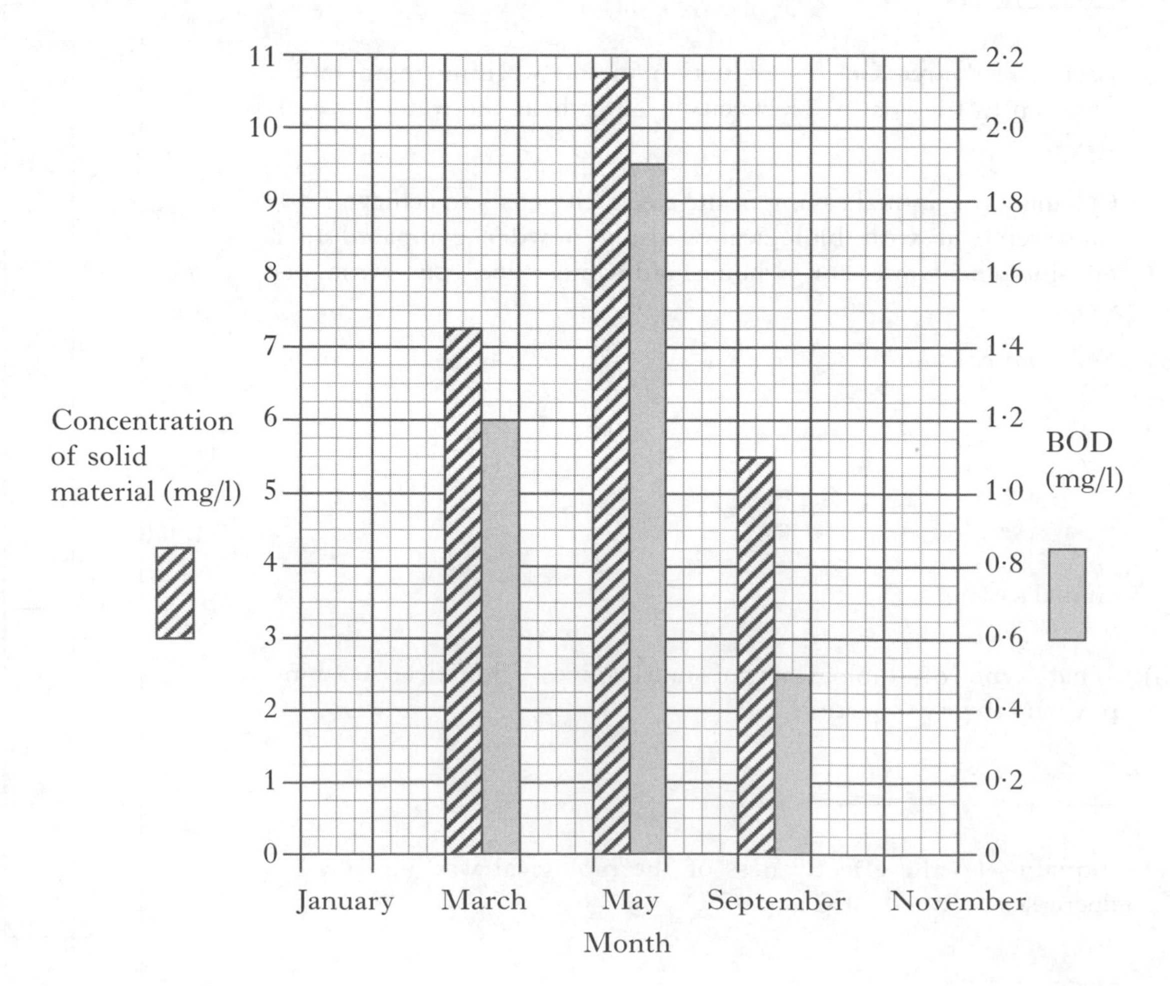

1

DO NOT WRITE IN THIS MARGIN

14. (continued)

Marks KU PS

(*b*) Describe the relationship between the concentration of solid material in the river water and the BOD.

____

____ **1**

(*c*) After heavy rains in December, the concentration of solid material in the water was found to be 10·0 mg/l.

What would be the expected BOD for this sample?
Tick (✓) the correct box.

7·5 mg/l ☐

5·0 mg/l ☐

1·75 mg/l ☐

1·25 mg/l ☐ **1**

[Turn over

DO NOT WRITE IN THIS MARGIN

Marks KU PS

15. Candytuft is a plant with white or pink flowers. The two forms of the gene responsible for the flower colour are:

P = pink flowers and **p** = white flowers.

(*a*) A plant breeder crossed two pink flowered plants as shown below.

Parents **Pp** × **Pp**

(i) What is the expected ratio of pink to white flowered plants in the offspring?

_____ : _____

pink : white

1

(ii) If 48 offspring had been produced, how many white flowered plants would have been expected?

Space for calculation

_____ white flowered plants

1

(iii) The offspring actually consisted of 24 pink flowered and 16 white flowered plants.

What is the simplest whole number ratio of pink to white flowered plants in the offspring?

Space for calculation

_____ : _____

pink : white

1

(iv) Suggest a reason for the difference between the expected ratio and the observed ratio.

____

DO NOT WRITE IN THIS MARGIN

Marks KU PS

15. (continued)

(*b*) What name is given to two different forms of a gene?

______________________ **1**

(*c*) Some plant characteristics show discontinuous variation. What is meant by "discontinuous variation"?

____

______________________ **1**

[*END OF QUESTION PAPER*]

DO NOT WRITE IN THIS MARGIN

KU	PS

ADDITIONAL GRAPH FOR QUESTION 10 (*b*) (iii)

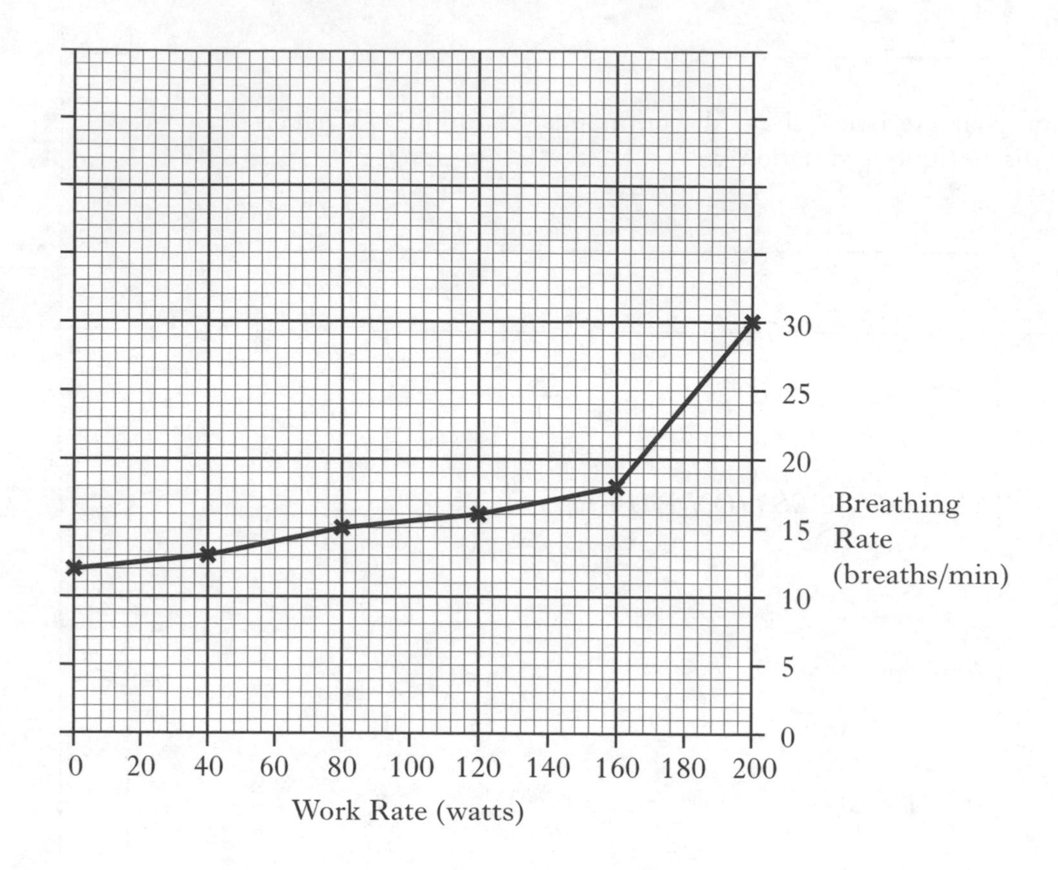

ADDITIONAL TIME LINE FOR QUESTION 12 (*d*)

Time line

Jan	Feb	Mar	Apr	May	Jun	Jul	Aug	Sep	Oct	Nov	Dec

DO NOT WRITE IN THIS MARGIN

KU	PS

ADDITIONAL CHART FOR QUESTION 14 (*a*)

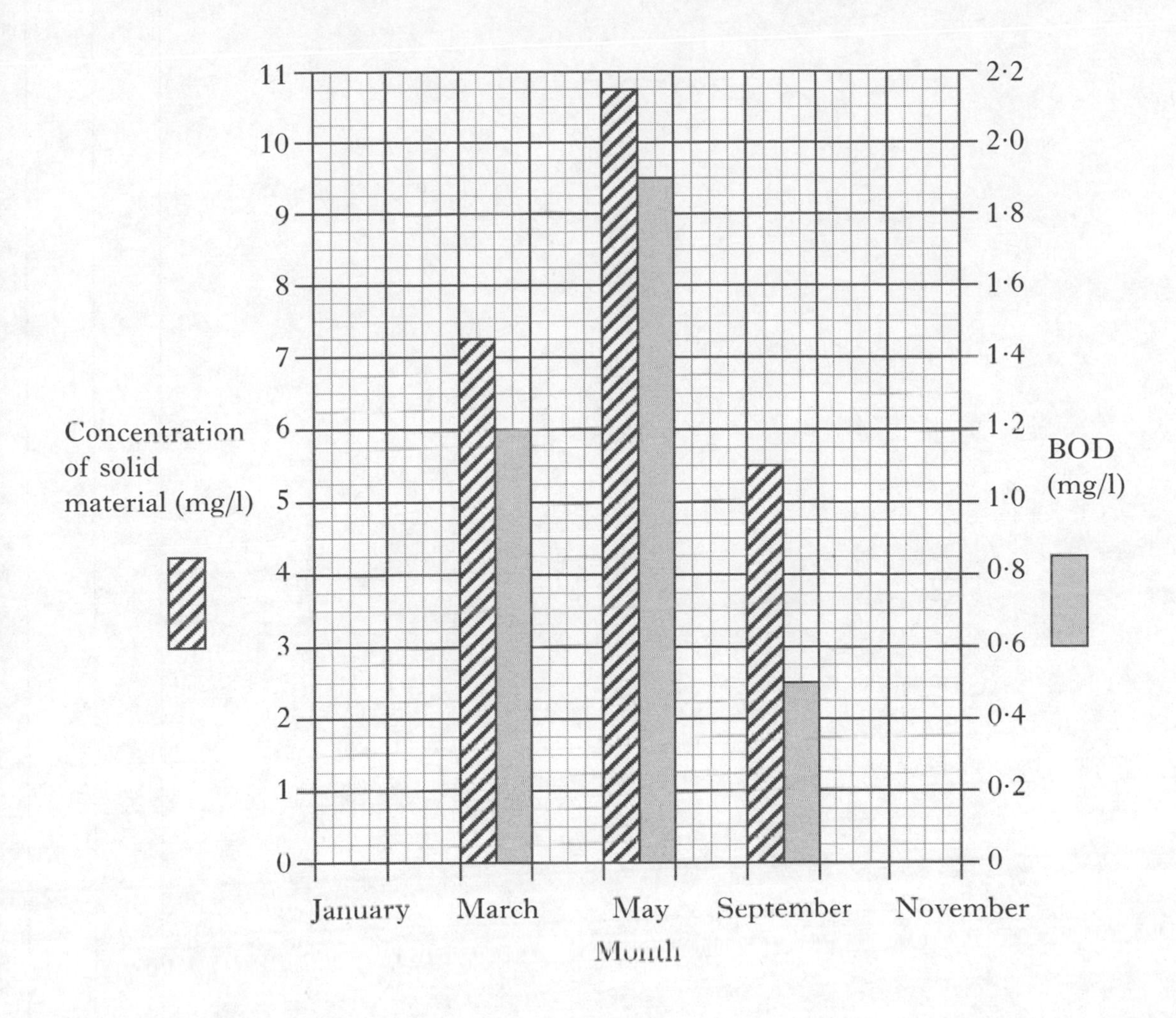

DO NOT WRITE IN THIS MARGIN

KU	PS

SPACE FOR ANSWERS
AND FOR ROUGH WORKING

STANDARD GRADE | CREDIT

2010

[BLANK PAGE]

FOR OFFICIAL USE

C

KU	PS

Total Marks

0300/402

NATIONAL QUALIFICATIONS 2010

THURSDAY, 27 MAY
10.50 AM – 12.20 PM

BIOLOGY STANDARD GRADE

Credit Level

Fill in these boxes and read what is printed below.

Full name of centre

Town

Forename(s)

Surname

Date of birth
Day Month Year

Scottish candidate number

Number of seat

1 All questions should be attempted.

2 The questions may be answered in any order but all answers are to be written in the spaces provided in this answer book, and must be written clearly and legibly in ink.

3 Rough work, if any should be necessary, as well as the fair copy, is to be written in this book. Additional spaces for answers and for rough work will be found at the end of the book. Rough work should be scored through when the fair copy has been written.

4 Before leaving the examination room you must give this book to the Invigilator. If you do not, you may lose all the marks for this paper.

DO NOT WRITE IN THIS MARGIN

Marks KU PS

1. (*a*) Two groups of pupils set pitfall traps in the school gardens to sample invertebrates living there. All traps were left for the same length of time. The results are shown in the following tables.

Group A	Pitfall trap number	Number of each type of invertebrate caught				
		spider	beetle	snail	earthworm	woodlouse
	1	2	1	2	0	1
	2	3	2	1	0	0

Group B	Pitfall trap number	Number of each type of invertebrate caught				
		spider	beetle	snail	earthworm	woodlouse
	1	2	3	2	1	1
	2	2	0	3	1	2
	3	0	2	1	1	1
	4	3	2	1	0	1
	5	3	1	1	2	1

(i) How many types of invertebrate did Group A find?

__________ types **1**

(ii) Calculate the average number of spiders found in Group B's traps.

Space for calculation

__________ spiders **1**

(iii) Explain why conclusions made by Group B from their results would be more reliable than conclusions made by Group A.

____

____ **1**

(iv) Give **one** precaution which must be taken when setting up a pitfall trap, or other named sampling technique, and explain the reason for it.

Sampling technique ______________________________

Precaution __________________________________

____ **1**

Reason ______________________________________

____ **1**

DO NOT WRITE IN THIS MARGIN

Marks KU PS

1. (continued)

(*b*) The diagrams below show the invertebrates collected by the pupils.

They are not drawn to scale.

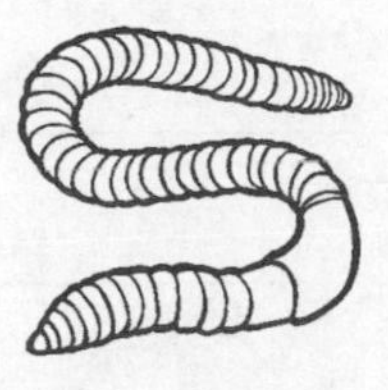
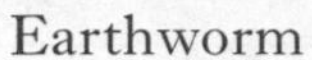
Earthworm

Snail

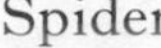
Spider

Beetle

Woodlouse

(i) Complete the following key using information from the diagrams.

1 Legs .... Go to 2

No legs .... Go to [ ] **1**

2 12 legs or more .... *Woodlouse*

Fewer than 12 legs .... Go to 3

3 Spots on body .... *Beetle*

No spots on body .... [ ] **1**

4 Shell .... *Snail*

[ ] .... [ ] **1**

(ii) Give **three** features of the beetle mentioned in the key.

1 ____

2 ____

3 ____ **1**

DO NOT WRITE IN THIS MARGIN

Marks KU PS

2. (*a*) Electricity can be generated by using fossil fuels or nuclear fuels as energy sources.

Give **one** disadvantage of using each type of fuel.

Fossil fuel ______________________________

______________________________ **1**

Nuclear fuel ______________________________

______________________________ **1**

(*b*) (i) Micro-organisms can obtain their energy by feeding on organic waste such as sewage.

Explain why each of the following events occurred after raw sewage was accidentally released into a river.

1 The number of micro-organisms in the river increased.

____

______________________________ **1**

2 The number of fish in the river decreased.

____

______________________________ **1**

(ii) A group of students monitored the river using indicator species.

What is meant by the term "indicator species"?

____

______________________________ **1**

DO NOT WRITE IN THIS MARGIN

Marks KU PS

3. (*a*) An investigation was carried out into the effect of temperature on the germination of grass seeds.

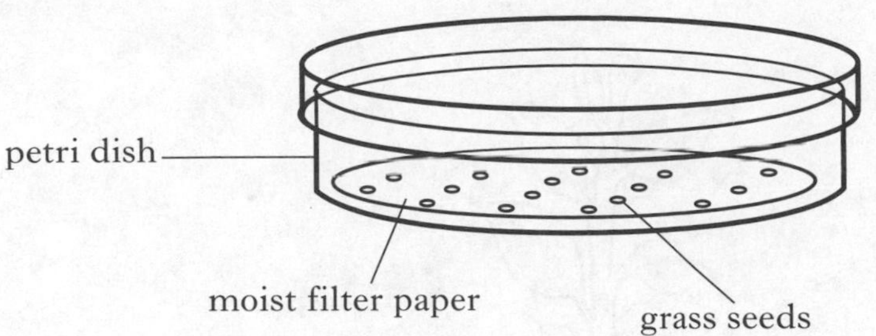

Five identical petri dishes, each containing 20 seeds, were set up as shown in the diagram. Each dish was left in the dark at a different temperature. After seven days the percentage germination in each dish was calculated. The results are shown in the table below.

Temperature (°C)	10	18	27	36	45
Percentage germination	45	65	80	70	40

(i) From the results, what is the optimum temperature for the germination of this species of grass?

_______ °C **1**

(ii) Name **one** factor, not already mentioned, which should be kept the same for all the dishes.

_______________________________ **1**

(iii) What feature of the investigation was designed to increase the reliability of the results?

_______________________________ **1**

(*b*) Describe the changes in the percentage germination of seeds that occur over a range of temperatures.

____

_______________________________ **2**

[Turn over

DO NOT WRITE IN THIS MARGIN

KU PS

4. Rooting compound helps plant cuttings to produce new roots. The diagram below shows the apparatus used to find out how the concentration of rooting compound affects this.

Six flasks were set up, each with a different concentration of rooting compound.

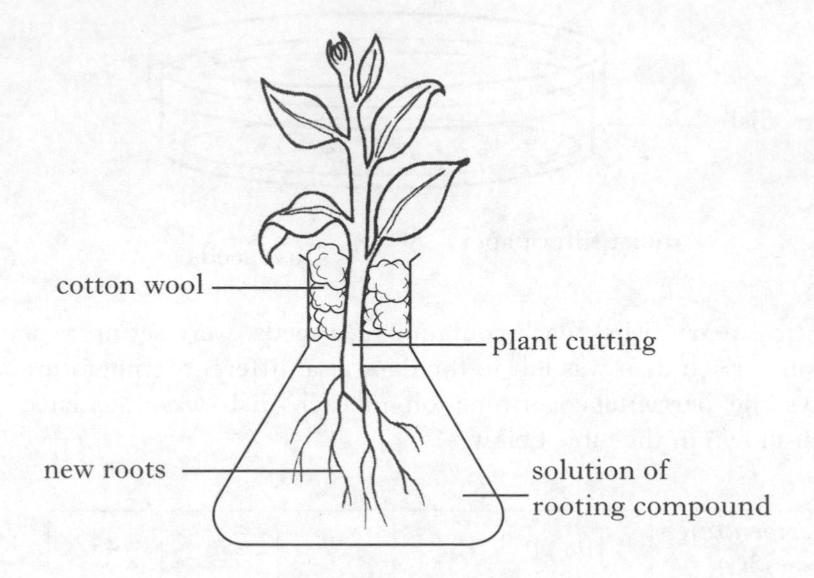

After 21 days the number of roots and the lengths of the roots on each cutting were measured.

The results are shown on the following graph.

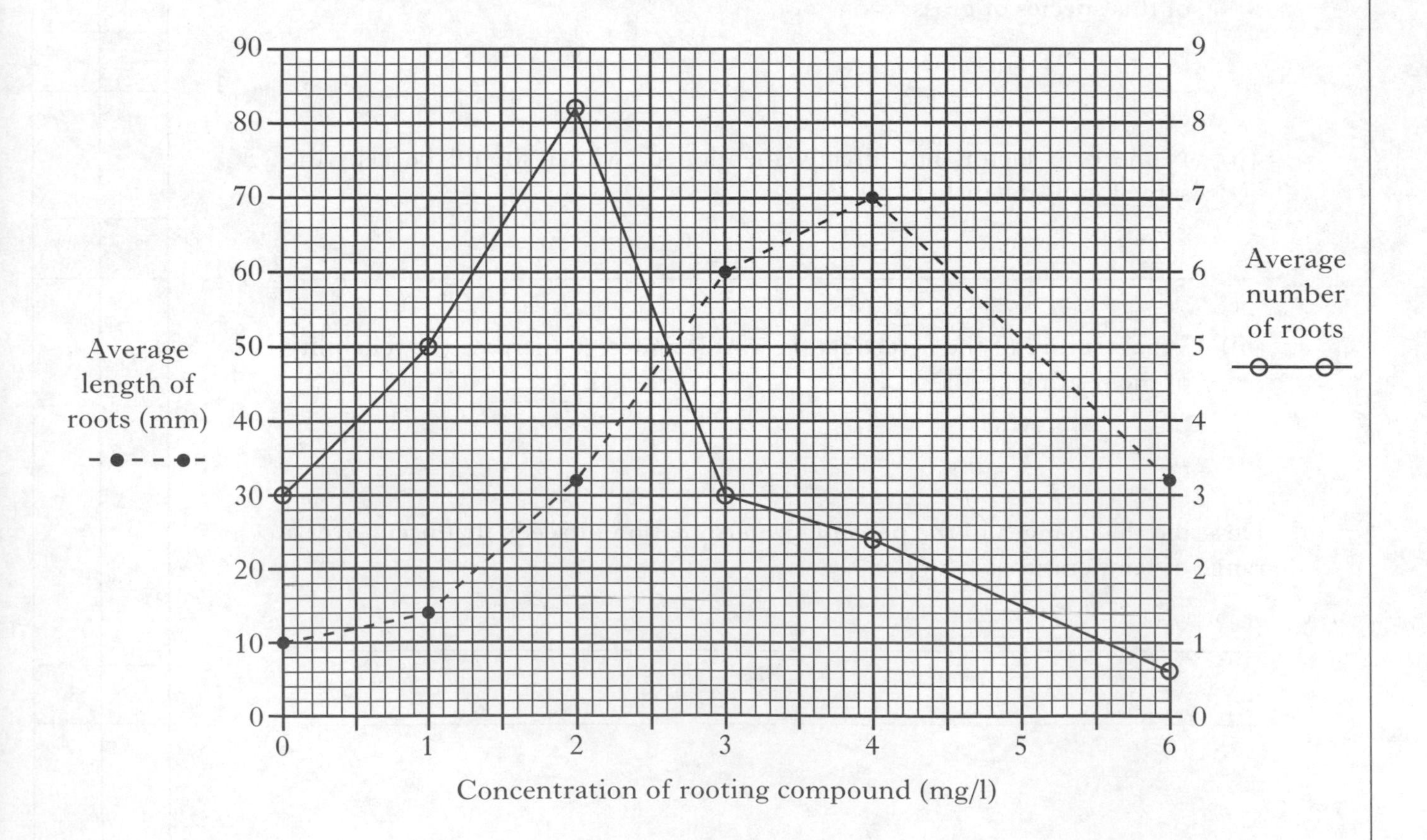

DO NOT WRITE IN THIS MARGIN

Marks KU PS

4. (continued)

(*a*) (i) Which **two** concentrations of rooting compound, used in the investigation, produced the same average root length?

__________ mg/l and __________ mg/l **1**

(ii) Using information from the graph, predict the average length of roots on cuttings grown in a concentration of 2·5mg/l.

__________ mm **1**

(iii) Which concentration of rooting compound produces the greatest number of roots per cutting?

__________ mg/l **1**

(iv) Describe how the average length of the roots on one cutting would be calculated.

____

____

____ **1**

(*b*) Give **one** advantage to a gardener of producing plants from cuttings rather than from seeds.

____ **1**

(*c*) What term is given to a group of plants grown from cuttings taken from a single plant?

________________________ **1**

[Turn over

DO NOT WRITE IN THIS MARGIN

Marks KU PS

5. (*a*) The following table gives information about reproduction in various animals.

	Average number of eggs or young produced per year	*Type of fertilisation*	*Where development takes place*
cod	6 million	external	water
frog		external	water
blackbird	5	internal	inside eggshell
stoat	4	internal	inside female

(i) A female frog produces a total of 4000 eggs over a five year period.

1 Complete the table to show the average number of eggs she produces per year.

Space for calculation

1

2 On average, two eggs from each female frog must survive to breeding age to keep the population constant. What percentage of this frog's **total** egg production does this represent?

Space for calculation

__________ %

1

(ii) Explain why fish such as cod must produce far more eggs than mammals such as stoats to ensure the survival of the species.

____

____ 1

(iii) Explain the importance of internal fertilisation to land-living animals.

____

____ 1

DO NOT WRITE IN THIS MARGIN

Marks | KU | PS

5. (continued)

(*b*) The diagram below represents a stage in the development of a human fetus.

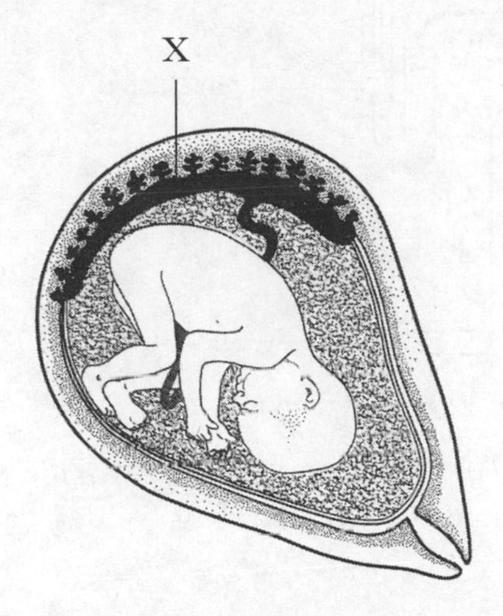

Name structure X and give **one** of its functions.

Name ______________________________

Function ______________________________

______________________________ 2

[Turn over

DO NOT WRITE IN THIS MARGIN

Marks KU PS

6. The apparatus shown below was used to study the effect of different temperatures on the activity of the enzyme catalase.

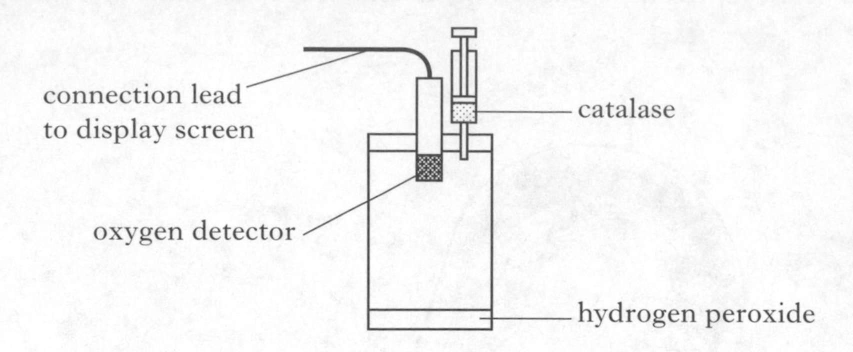

The catalase was added and reacted with the hydrogen peroxide to release oxygen. The increase in oxygen compared to the starting value was recorded as a percentage.

This was carried out at five different temperatures and the results are shown below.

Temperature (°C)	*Increase in oxygen* (%)
4	0·55
21	0·80
34	1·45
40	1·05
50	0·05

(*a*) Use the results to draw a line graph.

(An additional grid, if needed, will be found on *Page twenty-three*.)

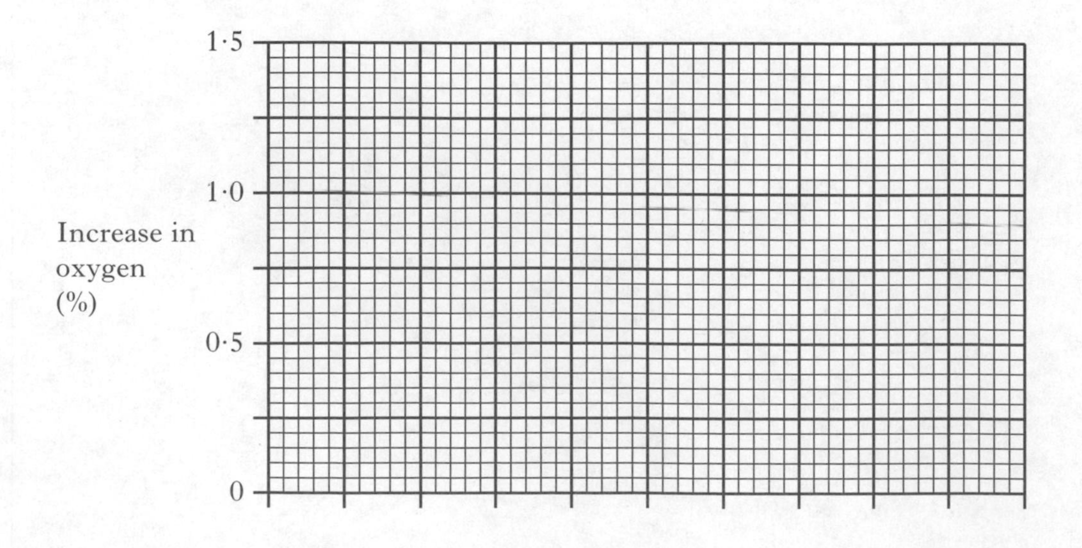

2

DO NOT WRITE IN THIS MARGIN

Marks | KU | PS

6. (continued)

(*b*) At which temperature was the catalase most active?

_________ °C — 1

(*c*) Why was it important that the catalase and the hydrogen peroxide were both at the required temperature before the catalase was added?

____

____ 1

(*d*) Explain why there was no oxygen released when the experiments were repeated with different enzymes.

____

____ 1

(*e*) Calculate the simple whole number ratio of percentage increase in oxygen at 34 °C, 40 °C and 50 °C.

Space for calculation

_____ : _____ : _____

34 °C 40 °C 50 °C — 1

[Turn over

DO NOT WRITE IN THIS MARGIN

Marks KU PS

7. The diagrams below represent red blood cells in different solutions as they would appear under a microscope.

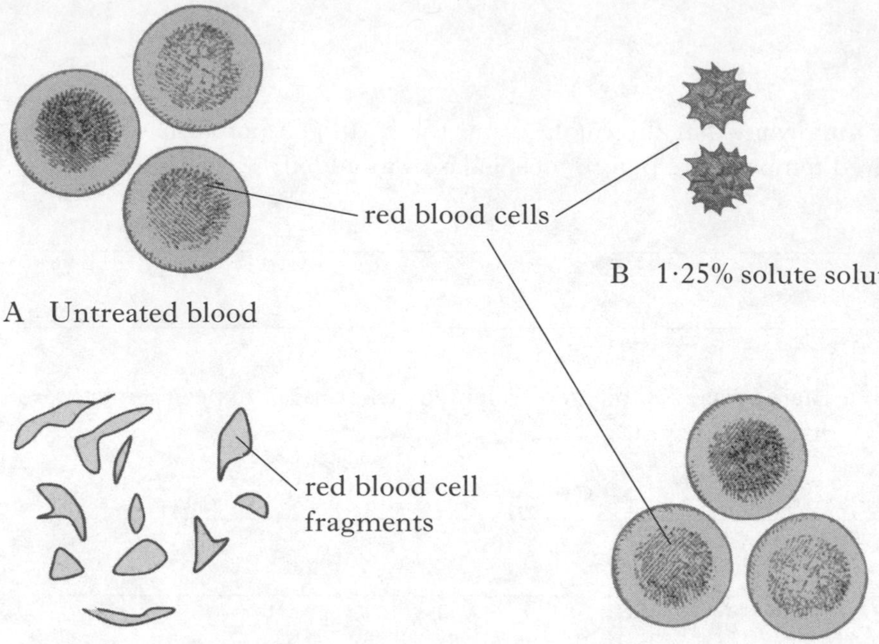

red blood cells

A Untreated blood

B 1·25% solute solution

red blood cell fragments

C 0·25% solute solution

D 0·90% solute solution

(*a*) Use the information in the diagrams to predict the percentage solute concentration of human blood. Explain your answer.

Solute concentration ________ %

Explanation ____

____

____ **1**

(*b*) What has happened to the cells in diagram B? Explain the change in terms of water concentrations.

Description ____

Explanation ____

____ **2**

DO NOT WRITE IN THIS MARGIN

Marks KU PS

8. The diagram below represents part of a finger joint.

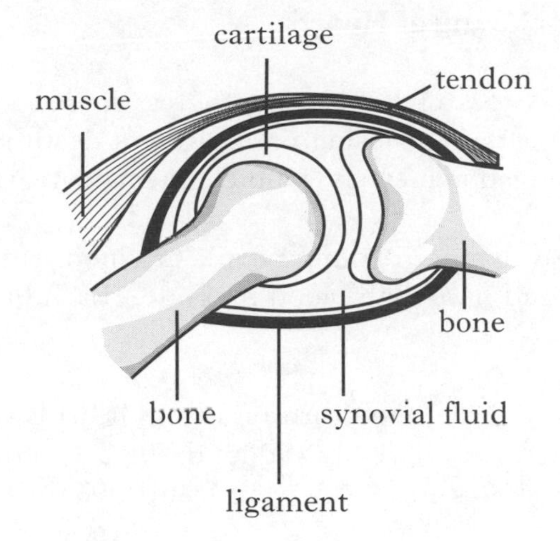

(*a*) (i) The joint needs a second muscle and tendon to make it function properly. Explain the need for joints to have muscles which work in pairs.

____

____

____ **2**

(ii) What feature of tendons ensures that all the force from a muscle contraction is transmitted to the bone?

____ **1**

(*b*) Name **two** parts of the joint which reduce friction

1 ________________________

2 ________________________ **1**

[Turn over

DO NOT WRITE IN THIS MARGIN

Marks KU PS

9. Read the following passage and answer the questions based on it.

Young at Heart?

New research shows that decades of hard-won progress in reducing the risk of heart disease in America appears to be losing pace. Recent death rates from heart disease remain almost unchanged in men and women under 55 years old.

This trend comes at a time when even young people are increasingly likely to be obese, suffer from diabetes and have high blood pressure. Each of these increases heart attack risk.

Data from 1980 to 2002 showed that the death rate from heart disease had fallen. In the whole population there was a yearly reduction of 2·9 percent during the 1980s, 2·6 percent during the 1990s and 4·4 percent from 2000 to 2002.

However the numbers told a strikingly different story for people aged 35 to 54. The yearly death rate from heart disease fell by 6·2 percent in the 1980s, by only 2·3 percent in the 1990s and showed no reduction at all between 2000 and 2002.

The message is that heart disease has not gone away, and could become an even greater problem if people fail to pay attention to known warning signs. Dr F S Ford, a medical officer for the American government said, "Young adults should take stock of their lifestyles. Don't smoke and take at least 30 minutes of exercise per day. If you need to lose weight, you must burn more energy than you take in. Good habits should start early. Changes that lead to heart disease, for example hardening of the arteries, occur at an early age. Therefore it is especially important that children and young people develop appropriate habits that minimise their risk of heart disease later in life."

(*a*) From the passage, identify **three** factors which contribute to the risk of heart disease.

1 ____________________

2 ____________________

3 ____________________ **1**

(*b*) Complete the table below to show the changes in death rates for the whole population and for the 35–54 age group.

	Average yearly reduction in death rate from heart disease (%)		
	1980–1989	1990–1999	2000–2002
Whole population			
35–54 age group			

2

DO NOT WRITE IN THIS MARGIN

Marks | KU | PS

9. (continued)

(*c*) According to Dr Ford, why is it important that "good habits should start early"?

____ **1**

(*d*) What cellular process is being referred to in the phrase "you must burn more energy"?

______________________________ **1**

[Turn over

DO NOT WRITE IN THIS MARGIN

Marks KU PS

10. A tin containing 170 g of evaporated milk has the following label.

Typical values per tin

Energy	1156 kJ
Protein	12·75 g
Carbohydrate	17·47 g
Fat	17·45 g
Fibre	0·00 g
Salt	0·33 g

(*a*) (i) What percentage of the total contents of the tin is protein?

Space for calculation

__________ % **1**

(ii) What component of the milk would provide most energy?

____________________ **1**

(*b*) Name the chemical elements present in fats.

____ **1**

DO NOT WRITE IN THIS MARGIN

Marks | KU | PS

11. (*a*) Underline **one** option in each bracket to make the following sentence about breathing correct.

When breathing out, the lung volume { increases / stays the same / decreases } and as a result the

air pressure in the lungs { increases / stays the same / decreases }. **1**

(*b*) The effect of changing the carbon dioxide concentration in inhaled air on a person's breathing was investigated.

The table below shows the average volume of air inhaled each minute at different concentrations of carbon dioxide.

Carbon dioxide concentration in inhaled air (%)	0	2	4	6	8
Average volume of air inhaled (litres per minute)	8	12	16	24	60

(i) How many times greater is the average volume of air inhaled per minute when the carbon dioxide concentration is increased from 2% to 8%?

Space for calculation

__________ times **1**

(ii) Calculate the average volume of carbon dioxide entering the lungs each minute when the carbon dioxide concentration in the air is 4%.

Space for calculation

__________ litres **1**

(iii) Calculate the increases in the average volume of air breathed per minute when the carbon dioxide changes from 0 to 2% and from 6 to 8%.

Express these increases as a simple whole number ratio.

Space for calculation

______ : ______

0–2% : 6–8% **1**

 [Turn over

DO NOT WRITE IN THIS MARGIN

Marks | KU | PS

12. (*a*) School pupils each carried out an identical word processing task. The resulting level of muscle fatigue was measured on a scale from 1 (low) to 7 (severe).

The results for the 95 pupils tested are shown in the following bar chart.

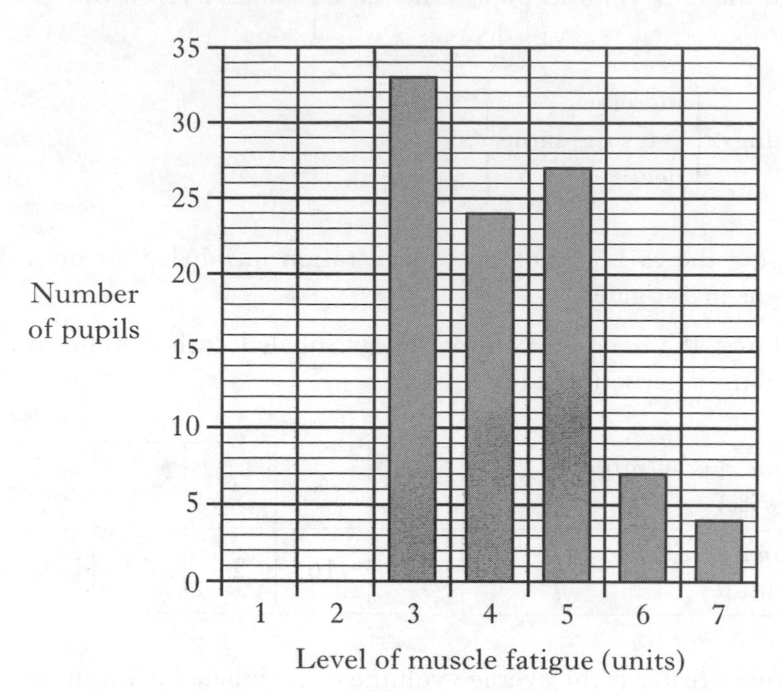

(i) Medical experts using this scale classify any score of 5 or more as "requiring urgent investigation". What percentage of the pupils tested were in this category?

Space for calculation

__________ % **1**

(ii) Give **two** conclusions which can be drawn from the results of this investigation.

1 ________________________________ **1**

2 ________________________________ **1**

(*b*) (i) What substance, produced by anaerobic respiration, causes muscle fatigue?

________________________ **1**

(ii) Explain why ensuring an adequate blood supply to muscles reduces the risk of muscle fatigue.

____

________________________________ **1**

DO NOT WRITE IN THIS MARGIN

Marks KU PS

13. The table below refers to egg production in the UK.

Living condition of hens	*Eggs laid* (percentage of total)
Living in cages	65
Living in barns	5
Free-range	30

(*a*) (i) Use the information from the table to complete the pie chart.

(An additional chart, if needed, will be found on *Page twenty-three*.)

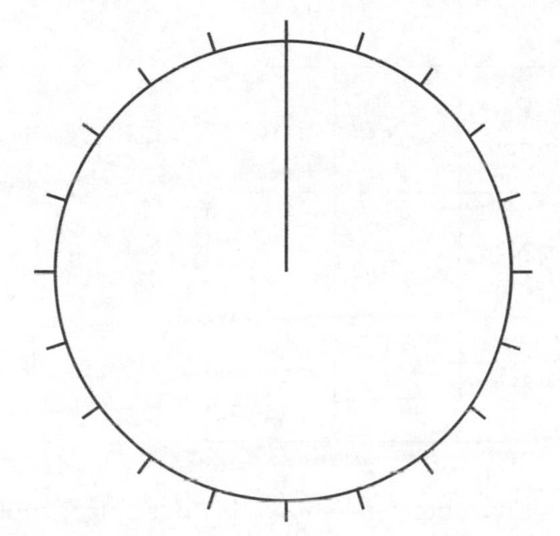

2

(ii) The total number of eggs laid per year in the UK is 30 million.

How many of these are laid by free-range hens?

Space for calculation

__________ eggs 1

(*b*) Modern varieties of hens can lay up to 300 eggs per year. Their ancestral wild varieties laid about 20 eggs per year.

(i) Calculate this increase in egg production as a percentage.

Space for calculation

__________ % 1

(ii) How has this improvement in egg production been achieved?

____ 1

DO NOT WRITE IN THIS MARGIN

Marks | KU | PS

14. Polydactyly is a condition which results in extra toes in mice. It is controlled by the dominant form of a gene (**N**). The normal phenotype is controlled by the recessive form (**n**).

The diagram below shows a cross between two mice of different genotypes.

Parent 1 **NN** × Parent 2 **nn**

↓

F_1 **Nn**

↓ allowed to interbreed

F_2

F_1 gametes	**N**	**n**
N		
n	**Nn**	

(*a*) (i) Complete the diagram above to show the possible genotypes of the F_2 generation. **1**

(ii) Give the phenotypes of each of the following mice.

Parent 1 ____________________

Parent 2 ____________________

F_1 ____________________ **2**

(iii) What term is used to describe the type of variation shown by these phenotypes?

____________________ **1**

(*b*) Why are the actual phenotype ratios in the F_2 generation often different from the predicted ones?

____

____________________ **1**

DO NOT WRITE IN THIS MARGIN

Marks KU PS

15. (*a*) Sucrose can be broken down into simple sugars using the enzyme invertase. The diagram below represents how this can be done commercially.

Sucrose solution is constantly being added and the products are constantly being removed.

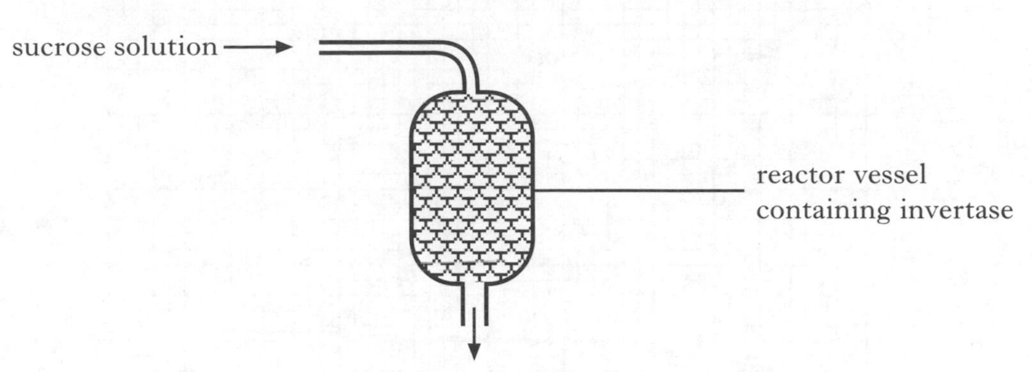

(i) What name is given to this type of process?

____________________ **1**

(ii) Explain why the enzyme does not leave the reactor vessel along with the products.

____________________ **1**

(*b*) (i) Genetic engineering techniques are used to produce enzymes which are used in biological washing powders. Which type of micro-organism is modified to produce the appropriate enzymes?

____________________ **1**

(ii) What is transferred from one organism to another during genetic engineering?

____________________ **1**

(*c*) During the brewing of beer, ingredients including yeast and malted barley are added to a fermentation vessel.

(i) What does the malted barley provide for fermentation which ungerminated barley does not?

____________________ **1**

(ii) How does sterilising the fermentation vessel before the raw materials are added help to provide optimum conditions for the yeast?

____________________ **1**

[Turn over for Question 16 on *Page twenty-two*

DO NOT WRITE IN THIS MARGIN

Marks KU PS

16. The concentrations of lactic acid and lactose in a milk sample were measured every two hours for 100 hours. The results are shown in the graph below.

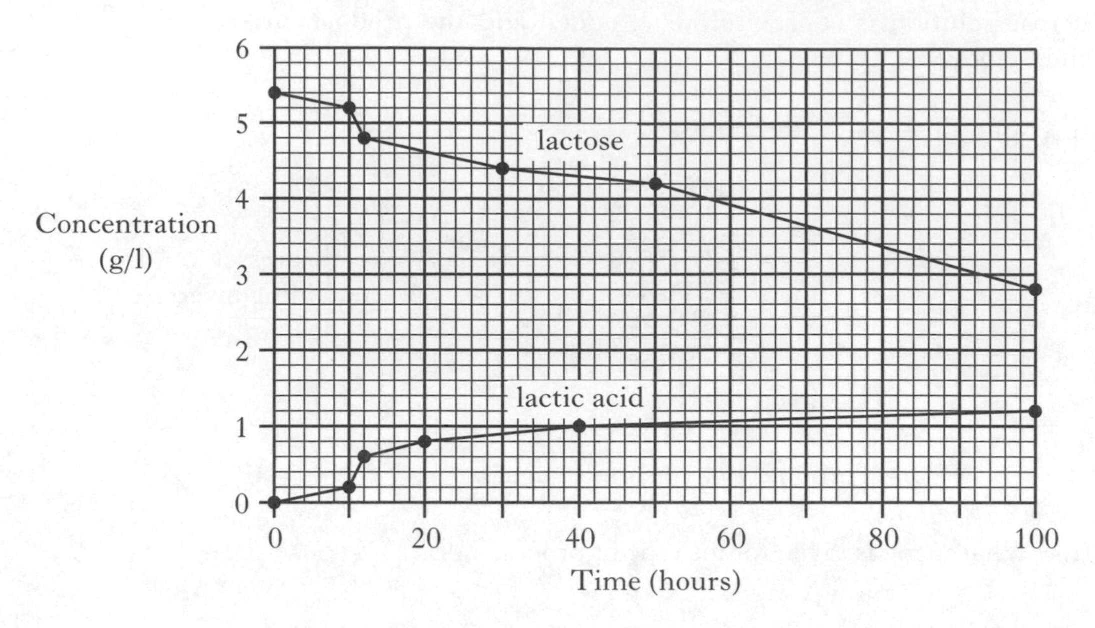

(*a*) (i) What evidence from the graph suggests that lactose is converted into lactic acid?

____

____ **1**

(ii) What evidence from the graph supports the theory that lactose is being converted into compounds other than lactic acid?

____

____ **1**

(*b*) Calculate the average hourly rate of lactose breakdown over the 100 hours of this investigation.

Space for calculation

__________ g/l/hour **1**

[*END OF QUESTION PAPER*]

SPACE FOR ANSWERS
AND FOR ROUGH WORKING

ADDITIONAL GRID FOR QUESTION 6(*a*)

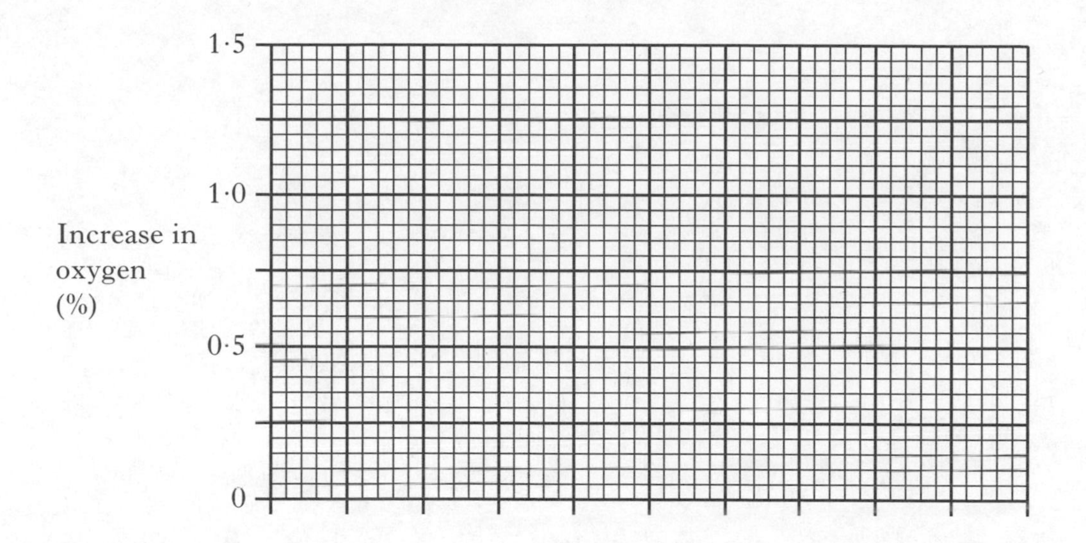

ADDITIONAL PIE CHART FOR QUESTION 13(*a*)(i)

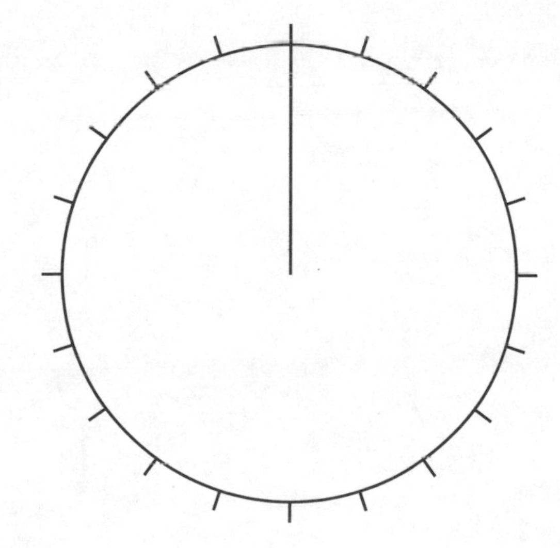

SPACE FOR ANSWERS
AND FOR ROUGH WORKING

[BLANK PAGE]

[BLANK PAGE]

[BLANK PAGE]

Acknowledgements

Permission has been sought from all relevant copyright holders and Bright Red Publishing is grateful for the use of the following:
The article 'North Sea is becoming too warm for cod and salmon' taken from The Herald, October 2003. Reproduced with permission of The Herald, Glasgow © 2009 Herald & Times Group (2006 page 13);
Article is adapted from 'Invasion of the Chinese Mitten Crab', from Biological Sciences Review, Volume 15, Number 2 (November 2002) published by Philip Allan Updates (2007 page 16);
An extract adapted from 'GM Organisms' by John Pickrell, taken from www.newscientist.com (2008 page 22);
An extract from 'The Life of Birds' by David Attenborough, published by BBC Books. Reprinted by permission of The Random House Group (2009 page 18);
An extract adapted from 'Coronary heart disease mortality among young adults in the US from 1980 through 2002. Concealed leveling of mortality rates' by Ford ES and Capewell S. Taken from J Am Coll Cardiol 2007; 50:2128–2132. Reproduced by permission of Dr ES Ford (2010 page 14).